T0197995

SOLVE IT!

COMPUTER-AIDED

MATHEMATICS

FOR

SCIENCE

AND

ENGINEERING

SOLVE IT!

COMPUTER-AIDED MATHEMATICS FOR SCIENCE AND ENGINEERING

SAMUEL DOUGHTY

Gulf Publishing Company
Houston, London, Paris, Zurich, Tokyo

SOLVE IT!

Computer-Aided Mathematics for Science and Engineering

Gulf Publishing Company
Book Division
P.O. Box 2608 □ Houston Texas 77252-2608

10 9 8 7 6 5 4 3 2 1

Printed in the United States of America

Library of Congress Cataloging-in-Publication Data
Doughty, Samuel.
    Solve it : computer-aided mathematics for science
and engineering / Samuel Doughty.
        p.   cm.
    Includes bibliographical references and index.
    ISBN 0-88415-266-9 (alk. paper)
    1. Engineering mathematics—Data processing.
    2. Science—Mathematics—Data processing.
    TA345.D68   1995
    519.4′0285′51—dc20   95-30439
                                                      CIP

Printed on Acid-Free Paper (∞).

For Ann-Marie, John, and Jenni

Contents

Foreword

The landscape of applied mathematics has been greatly changed by the computer, and more particularly, by the microcomputer that is found on virtually every desk. Before the computer, many solution techniques led to *formal solutions*, solutions that showed the form of the result but often only indicated some major computational effort required to actually evaluate the result. The implication was, "Here is what it will look like and what is required, but it will only be usable if you are willing to do this rather substantial amount of computation to implement it." Today, that "rather substantial amount of computation" is often readily done with the computer.

Even with the availability of the computer, there is still the need for detailed computation instructions known as *algorithms*. For example, many formal solutions involve an indicated matrix inverse, but the details of the matrix inversion are not specified. There are numerous methods for matrix inversion, each preferred for a particular situation. Before the formal solution can be turned into a numerical solution, a specific method of inversion must be selected and the detailed computational steps must be specified, that is, an algorithm must be established.

This book is not intended to be a comprehensive book on numerical analysis. Rather, it is intended to be a companion, a knowledgeable friend, offering help in those areas where it has experience. That experience comes from the writer's approximately thirty years of numerical solutions for problems in science and engineering. The topics included are those the writer has found to be useful, but which are not readily available in a single source. These matters range from topics that will be useful to a college freshman, such as curve fitting, to topics more likely to useful to the practicing professional, such as applications of Green's theorem.

It is the author's hope that this book will become a friend of the user early in his college career and remain with the user into his professional life. The term *user* is employed, rather than *reader*, because it is the author's intent that this book be *used*, not simply read. The book is written with this purpose in mind, and it frequently includes sections of sample computer code, often subroutines or main programs, which the user is encouraged to first program as they stand, then check, debug, and adapt the program to the user's problems.

The author's intent has been to present the topics as independent units. The exception is the discussion on matrices and matrix notation that appears quite early. This material is used rather freely throughout the remainder of the book, and the user will do well to read and understand the use of matrices to this extent before plunging into the remainder of the book.

Considerable effort has been expended to eliminate errors, but no doubt some remain. The author will be grateful to those users who pass on corrections, comments, and suggestions for improvement.

Samuel Doughty

SOLVE IT!

COMPUTER-AIDED

MATHEMATICS

FOR

SCIENCE

AND

ENGINEERING

Chapter 1

Philosophy

Make It Easy For Whom?

There are many reasons to use digital computation today including labor reduction, elimination of random, careless mistakes, and increased accuracy. The major reason, in most cases is the first one, labor reduction. This includes the opportunity to use methods that would be simply too tedious for manual computation and are only accessible through the digital computer. If labor reduction is the primary concern, then it is important to assure that human labor is really reduced. All too often, human effort is expended to make the work easy for the machine, and this is usually contrary to the goal of reducing human labor.

Some people will make the elementary mistake of trying to save effort in the execution of the program by precomputing various constants to be inserted into the code directly. A common example of this is the calculation of the value of 2π and the programming of a decimal approximation to this value. If the programming language in use has an internal value for π, then the reasonable approach is to let the machine make the multiplication by 2 whenever the program is executed. This will give a value with as many significant figures as are being calculated within the program. If the language does not have an internal value for π, then such a value should be programmed once, near the beginning of the code, and used by name thereafter. The repeated programming of a decimal approximation is both tedious and a frequent source of errors.

A more subtle waste of time occurs in seeking more and more complicated programming techniques, usually for reduced execution time. In many technical applications, the final program will be executed only once or a few times to obtain the desired results, and this is the sort of computation discussed here. For applications such a real-time control, there is clearly a need for minimum execution time, but that is outside the scope of this book. There is obviously a trade-off to be made here, because it is human time that is often wasted while waiting for execution to be completed. Even so, there is little gained by spending an extra hour of programming time in order to reduce execution time by a few milliseconds total. Execution time becomes a problem when it is measured in hours or even days, but relatively few programs approach such times.

A second waste of time often involved with more complicated programming comes from the fact that the resulting code is more difficult to debug. By developing complicated code, particularly long, complicated individual statements, the time required to search out errors is often multiplied many times.

Write It Yourself or Go Commercial?

There is a dazzling array of commercial software available today to do everything from managing your social calendar to arranging musical scores. The software vendors would like for you to believe that their product is superior to anything that you could write, and that to do a satisfactory job, you have no choice but to use their product. For a computing problem for which you do not understand the process, they may very well be correct. For any task that you understand how to do well, there is often no need to go to a commercial program. There certainly is no need to look for a commercial code for any sort of calculation where you are the expert and they are the amateurs!

One of the principal advantages of writing the code yourself is that you know exactly what it does, exactly what assumptions are built into the process. If there is a need to revise the solution, you can certainly do so if you wrote the code in the first place. If you are using a commercial code and need to make a revision, there is really very little that you can do other than to petition the software vendor for a revision or look for another vendor. The security of knowing exactly how the program works, and the associated flexibility to make changes as needed, often outweigh the short term advantages of using a commercial code, not to mention the cost of the commercial code. And remember, commercial codes cost twice: once for the software and again for the time and effort required to develop proficiency!

Programming Language

If you are going to write the code yourself, you must chose a programming language. The choice of programming language is often dictated by what is available. For most scientific and engineering computation (not real-time applications), a language that facilitates program development is the best choice. Depending on the nature of the work being programmed, features such as matrix operations and high level plotting commands are often significant.

Most microcomputers offer a version of BASIC, and this language lends itself to program development. There are, however, many different versions of BASIC, and they are not all equal. One of the most powerful versions is True BASIC, a product of True BASIC, Inc., West Lebanon, NH. True BASIC includes matrix operations and high level plotting commands, and execution is fairly fast on most problems. The author has found this language to have an optimum combination of programming commands, speed of execution, and ease of usage for most of his engineering consulting work. This is the language used for all of the examples in this book.

Chapter   2

Matrices

A user of this book might wonder why the opening discussion is on a topic as abstract as matrices. The reasons for this choice are

- although often presented abstractly, matrices are very useful for a wide variety of applied mathematics used in science and engineering, and

- matrix notation is quite close to what is needed for computer program development in many situations. Consequently, a problem formulated in matrix notation is usually rather easily programmed for computer numerical solution.

With these two thoughts in mind, matrices are a very good place to begin this presentation. It should be emphasized that very little matrix theory will be used; the emphasis is primarily on the notation and arithmetic of matrices.

Matrix Notation

The terms "matrix" or "matrix notation" as used in applied mathematics refer to a rectangular array of numbers. The word *matrix* is singular; to refer to more than one matrix, the ending changes to become *matrices*. Thus it is possible to have *one matrix* that is the sum of *two matrices*. The numbers in the array are called *elements* of the matrix, and the elements are arranged in rows and columns. A typical matrix, $[A]$ with n_r rows and n_c columns represents the following array:

$$[A] = \begin{bmatrix} a_{11} & a_{12} & a_{13} & \cdots & a_{1n_c} \\ a_{21} & a_{22} & a_{23} & \cdots & a_{2n_c} \\ a_{31} & a_{32} & a_{33} & \cdots & a_{3n_c} \\ \vdots & \vdots & \vdots & \ddots & \\ a_{n_r 1} & a_{n_r 2} & a_{n_r 3} & & a_{n_r n_c} \end{bmatrix}$$

The typical element is denoted as a_{ij} where the first subscript is the *row index* and the second subscript is the *column index*. In this way, the indices on each element indicate its location within the array. All of the elements in the first row, have a 1 as the first index. Similarly, all of the elements in the second row have a 2 as the first index, etc. All of the elements in the first column have a 1 as the second index. All of the elements in the second column have a 2 for the second index, etc. The range of the row index is from 1 to n_r and is called the *row dimension* of the array. The range of the column index is from 1 to n_c, and this is called the *column dimension* of the array.

We are interested in two types of matrices. The first is called a square matrix. A matrix is square, provided that the row and column dimensions are equal, $n_r = n_c$. It is spoken of as an $(n \times n)$ array. In form, a square matrix looks just like the matrix $[A]$ above and requires no further comment at this point. The second form of interest is called a *column (or row) vector*. For a column vector, the column dimension is 1, so the array has the form

$$\{B\} = \left\{ \begin{array}{c} b_1 \\ b_2 \\ \vdots \\ b_n \end{array} \right\}$$

The braces, { }, are used to denote a column vector, in contrast with the brackets [ ] used above for a rectangular or square matrix. Because it is a one dimensional array, only a single index is required and there is only a single dimension, a row dimension. This same array can be written in row form as

$$(B) = (b_1, b_2, \cdots b_n)$$

where the parentheses ( ) denote the row format.

If the idea of a matrix as a two dimensional array is to be maintained, then the column vector should be considered as a $(n \times 1)$ array. Similarly, the row vector is a $(1 \times n)$ array. The value of this formal consistency with the rectangular array description will become more evident shortly.

One of the basic matrix operations is called the *transpose*. The transpose consists of interchanging the rows and columns in a rectangular matrix so that the elements comprising a row in the original matrix become a column in the transposed matrix. The transpose of the matrix $[A]$ above, written with indices that denote position in the original array, is

$$[A]^t = \begin{bmatrix} a_{11} & a_{21} & a_{31} & \cdots & a_{n_r 1} \\ a_{12} & a_{22} & a_{32} & \cdots & a_{n_r 2} \\ a_{13} & a_{23} & a_{33} & \cdots & a_{n_r 3} \\ \vdots & \vdots & \vdots & \ddots & \\ a_{1n_c} & a_{2n_c} & a_{3n_c} & & a_{n_c n_r} \end{bmatrix}$$

The superscript $\mathbf{t}$ on the matrix $[A]$ is not an exponent, but rather is the notation for the transpose of the matrix. When the transpose operation is applied to a column vector, the result is the row vector form for the same array, thus:

$$\{B\}^t = \left\{ \begin{array}{c} b_1 \\ b_2 \\ \vdots \\ b_n \end{array} \right\}^t = (b_1, b_2, \cdots b_n) = (B)$$

and similarly, the row form is converted to the column form by transposition:

$$(B)^t = (b_1, b_2, \cdots b_n)^t = \left\{ \begin{array}{c} b_1 \\ b_2 \\ \vdots \\ b_n \end{array} \right\} = \{B\}$$

Note that $[A]^{t^t} = [A]$, that is, two applications of the transpose operation simply restore a matrix to its original form.

At this point, it is appropriate to define two special matrix forms. A diagonal matrix is a square matrix with all elements zero, except for those on the *main diagonal*, which is the group of elements lying along the diagonal from the upper left corner to the lower right corner. Since the only nonzero values are found on the main diagonal, only those values need to be expressed if it is known that a particular matrix is diagonal. This is often written as

$$[D] = Diag\,(d_1, d_2, \cdots d_n)$$

where the notation $Diag\,(\cdots)$ conveys the fact that the matrix is diagonal and the d_i are the values of the diagonal elements.

The second definition required is that of the *identity* matrix. An identity matrix is a square, diagonal matrix with unit values on the main diagonal. Thus, for a dimension of 3, the identity matrix is

$$[I] = Diag\,(1, 1, 1) = \begin{bmatrix} 1 & 0 & 0 \\ 0 & 1 & 0 \\ 0 & 0 & 1 \end{bmatrix}$$

The notation $[I]$ is widely used for the identity matrix, but caution should be used where this may be confused with other matrices such as the inertia matrix, which is often denoted by the same symbol. The identity matrix is in some ways comparable to a 1 in scalar arithmetic; it is the unit element in matrix arithmetic.

Addition and Subtraction of Matrices

Either addition or subtraction of two matrices produces a resultant matrix where each element of the result is the sum or difference of the corresponding terms in the two matrices being added or subtracted. Such a description makes sense only when the two original matrices have the same row and column dimensions. Further, the resultant matrix will have the same row and column dimensions as the two original matrices. If $[A]$ and $[B]$ are rectangular arrays having dimensions $(n_r \times n_c)$,

$$[A] = \begin{bmatrix} a_{11} & a_{12} & \cdots \\ a_{21} & a_{22} & \\ \vdots & & \ddots \end{bmatrix} \qquad [B] = \begin{bmatrix} b_{11} & b_{12} & \cdots \\ b_{21} & b_{22} & \\ \vdots & & \ddots \end{bmatrix}$$

the sum and difference are $[S]$ and $[D]$, respectively,

$$[S] = \begin{bmatrix} s_{11} & s_{12} & \cdots \\ s_{21} & s_{22} & \cdots \\ \vdots & \vdots & \ddots \end{bmatrix} = \begin{bmatrix} a_{11} + b_{11} & a_{12} + b_{12} & \cdots \\ a_{21} + b_{21} & a_{22} + b_{22} & \cdots \\ \vdots & \vdots & \ddots \end{bmatrix}$$

$$[D] = \begin{bmatrix} d_{11} & d_{12} & \cdots \\ d_{21} & d_{22} & \cdots \\ \vdots & \vdots & \ddots \end{bmatrix} = \begin{bmatrix} a_{11} - b_{11} & a_{12} - b_{12} & \cdots \\ a_{21} - b_{21} & a_{22} - b_{22} & \cdots \\ \vdots & \vdots & \ddots \end{bmatrix}$$

Multiplication of Matrices

If two matrices are multiplied together, the result is another matrix. The elements of the result are the sums of products of the elements of the two matrix factors. To be more specific, and using the notation for the two matrices $[A]$ and $[B]$ defined above, the elements of the product matrix, $[P]$, are computed as:

$$p_{ik} = \sum_j a_{ij} b_{jk}$$

If the sum on j is to make sense, the range of values for j in the $[A]$ matrix (the column dimension of $[A]$) must be the same as the range of values for j in the $[B]$ matrix (the row dimension of $[B]$). Evidently i can take all of the values it was allowed in the $[A]$ matrix, while k can assume all values that were allowed for it in the $[B]$ matrix. The result is a square matrix having the row dimension from the first factor and the column dimension from the second factor:

$$\underbrace{[P]}_{n_1 \times n_3} = \underbrace{[A]}_{n_1 \times n_2} \underbrace{[B]}_{n_2 \times n_3}$$

Note that this notation indicates a common value for the summed index, n_2, as shown. It also shows the row dimension of the product as the row dimension of the first factor and the column dimension of the product as the column dimension of the second factor. If any of this condition is not met, the matrix product is not defined.

It is necessary to point out that matrix multiplication is *noncommutative*, which is to say that the order of the factors is significant. This means that, in general $[A][B] \neq [B][A]$, even if both products are properly defined. There may be particular instances in which the order of the factors may be reversed, but this must be verified for the particular situation. One example of a case where matrices do commute is found when the identity matrix is one of the factors:

$$[A][I] = [I][A] = [A]$$

Multiplying by the identity matrix is comparable to scalar multiplication by 1.

It will often be useful to multiply matrices within a computer program. The following code segment, based on the expression for the typical term given above, illustrates how this is done.

Example Code for Matrix Product

```
for i = 1 to n1
   for k = 1 to n3
      sum = 0
      for j = 1 to n2
         sum = sum + a(i,j) * b(j,k)
      next j
   next k
 next i
```

This example code assumes that the arrays a(,), b(,) and c(,) have been previously defined and that n1, n2, and n3 have the appropriate values to match the array dimension statements.

Matrix Inverse

Thus far, matrices appear to behave much like scalars, at least for addition, subtraction, and multiplication operations. (The notable exception is that matrix multiplication does not commute whereas scalar multiplication is commutative.) There is, however, no division operation for matrices, so the analogy fails at this point. For many matrices there is a second matrix said to be the *inverse* of the first, but this is not equivalent to a scalar reciprocal.

Assume that $[A]$ is a square matrix having an inverse (more will be said later about when an inverse exists). *If the matrix product* $[A][V]$ *results in the identity matrix, then* $[V]$ *is called the inverse of* $[A]$. The inverse of $[A]$ is often written as $[A]^{-1}$, where the superscript -1 suggests the exponent -1 denoting a reciprocal in scalar arithmetic. While some would prefer to distinguish between a left inverse satisfying $[A]^{-1}[A] = [I]$ and a right inverse satisfying $[A][A]^{-1} = [I]$, as a practical matter and for computational purposes, they are the same. Thus,

$$[A][A]^{-1} = [A]^{-1}[A] = [I]$$

The analogy with scalar arithmetic is obvious, but note again that there is no division operation defined for matrices, only multiplication. This is true, even when one of the factors in the product is described as an "inverse."

Inverse matrices exist only for square matrices, but not every square matrix has an inverse. A matrix for which there is no inverse is called *singular*, meaning exceptional or out of the ordinary. A later section will deal further with the question of when an inverse exists and what failure to have an inverse indicates about the matrix. While the description above provides a means by which a possible inverse matrix can be tested, it does not really provide a means for determining the inverse if it exists.

This matter is closely bound up with the solution of systems of simultaneous, linear algebraic equations, and it will be presented in the following section.

Solution of Systems of Linear Equations

Many problems involving physical systems result in systems of linear, simultaneous, algebraic equations of the form

$$a_{11}x_1 + a_{12}x_2 + a_{13}x_3 + \cdots = c_1$$
$$a_{21}x_1 + a_{22}x_2 + a_{23}x_3 + \cdots = c_2$$
$$\vdots$$
$$a_{n1}x_x + a_{n2}x_2 + a_{n3}x_3 + \cdots = c_n$$

Many problems can be cast in this form with the appropriate definitions, even though the initial formulation would rarely use such strange notation (lots of doubly subscripted c's, all of the unknowns denoted as subscripted x's, and right side values all denoted by b with numerical subscripts). In terms of matrix notation, this system can be written as

$$\begin{bmatrix} a_{11} & a_{12} & \cdots & a_{1n} \\ a_{21} & a_{22} & \cdots & a_{2n} \\ \vdots & & \ddots & \vdots \\ a_{n1} & a_{n2} & \cdots & a_{nn} \end{bmatrix} \begin{Bmatrix} x_1 \\ x_2 \\ \vdots \\ x_n \end{Bmatrix} = \begin{Bmatrix} c_1 \\ c_2 \\ \vdots \\ c_n \end{Bmatrix}$$

or, more compactly, just

$$[A]\{X\} = \{C\}$$

The formal solution for this system is accomplished by premultiplying (multiplying on the left) by the inverse of $[A]$:

$$[A]^{-1}[A]\{X\} = [A]^{-1}\{C\}$$
$$[I]\{X\} = [A]^{-1}\{C\}$$
$$\{X\} = [A]^{-1}\{C\}$$

If the matrix $[A]^{-1}$ can be determined, then the system can be solved as shown above. Conversely, if a solution can be found for the system of equations, then in a manner of speaking, the inverse will have been determined.

The nineteenth century German mathematician Gauss devised a systematic approach to the solution for systems of linear algebraic equations that is presented below. Rather than develop the process in completely general notation, a specific example will be used, recognizing that the application to any other system follows the same general lines. Consider the system of three equations:

$$a_{11}x_1 + a_{12}x_2 + a_{13}x_3 = c_1$$
$$a_{21}x_1 + a_{22}x_2 + a_{23}x_3 = c_2$$
$$a_{31}x_1 + a_{32}x_2 + a_{33}x_3 = c_3$$

For the operations required it is convenient to deal only with the coefficients and the right side elements, in a form known as the *augmented matrix of coefficients*, that is the matrix of coefficients, $[A]$, augmented or expanded to include the right side elements as well:

$$\begin{bmatrix} a_{11} & a_{12} & a_{13} & \mid & c_1 \\ a_{21} & a_{22} & a_{23} & \mid & c_2 \\ a_{31} & a_{32} & a_{33} & \mid & c_3 \end{bmatrix}$$

Now suppose that an appropriate multiple of the first row is added to the second row such as to produce a zero in the $(2,1)$ position. The required multiple is $-a_{21}/a_{11}$. In a similar manner, adding $-a_{31}/a_{11}$ times the first row to the third row will produce a zero in the $(3,1)$ location. The result will be

$$\begin{bmatrix} a_{11} & a_{12} & a_{13} & \mid & c_1 \\ a_{21} + \left(\frac{-a_{21}}{a_{11}}\right) a_{11} & a_{22} + \left(\frac{-a_{21}}{a_{11}}\right) a_{12} & a_{23} + \left(\frac{-a_{21}}{a_{11}}\right) a_{13} & \mid & c_2 + \left(\frac{-a_{21}}{a_{11}}\right) c_1 \\ a_{31} + \left(\frac{-a_{31}}{a_{11}}\right) a_{11} & a_{32} + \left(\frac{-a_{31}}{a_{11}}\right) a_{12} & a_{33} + \left(\frac{-a_{31}}{a_{11}}\right) a_{13} & \mid & c_3 + \left(\frac{-a_{31}}{a_{11}}\right) c_1 \end{bmatrix}$$

or, using a prime $(')$ to denote a modified value, this can be written as

$$\begin{bmatrix} a_{11} & a_{12} & a_{13} & \mid & c_1 \\ 0 & a'_{22} & a'_{23} & \mid & c'_2 \\ 0 & a'_{32} & a'_{33} & \mid & c'_3 \end{bmatrix}$$

Next, a multiple of the second row will be added to the third row in such a way as to induce a zero in the $(3,2)$ position. This is done leaving the first and second rows unchanged. The necessary multiple is $-a'_{32}/a'_{32}$:

$$\begin{bmatrix} a_{11} & a_{12} & a_{13} & \mid & c_1 \\ 0 & a'_{22} & a'_{23} & \mid & c'_2 \\ 0 & a'_{32} + \left(\frac{-a'_{32}}{a'_{22}}\right) a'_{22} & a'_{33} + \left(\frac{-a'_{32}}{c'_{22}}\right) a'_{23} & \mid & c'_3 + \left(\frac{-a'_{32}}{a'_{22}}\right) c'_2 \end{bmatrix}$$

or,

$$\begin{bmatrix} a_{11} & a_{12} & a_{13} & \mid & c_1 \\ 0 & a'_{22} & a'_{23} & \mid & c'_2 \\ 0 & 0 & a''_{33} & \mid & c''_3 \end{bmatrix}$$

where the double prime denotes yet another revision of the element value.

If the last line is recast in scalar form, it is simply

$$a''_{33} x_3 = c''_3$$

a relation easily solved for x_3. With a value now known for x_3, the second equation can be used to evaluate x_2, and finally the first equation is used to evaluate x_1:

$$x_3 = \frac{c''_3}{a''_{33}}$$
$$x_2 = \frac{c'_2}{a'_{22}} - \frac{a'_{23}}{a'_{22}} x_3$$
$$x_1 = \frac{c_1}{a_{11}} - \frac{a_{12}}{a_{11}} x_2 - \frac{a_{13}}{a_{11}} x_3$$

For rather obvious reasons, this last phase is often called "back substitution." This is the essence of the method of *Gaussian elimination* as it is commonly called; the name derives from the fact that the unknowns were systematically eliminated one by one from the equations. It is a *back* substitution in the sense that it proceeds in the direction opposite to that which introduced the zeroes into the coefficient matrix.

One other point needs to be mentioned with regard to the numerical implementation of the approach. As seen above, the coefficient a_{11} appeared as divisor in many fractions. It would be desirable for this number to be as large as possible in order to minimize round-off errors in the subsequent difference operation. There is no significance to the order of the equations; they may be freely interchanged. Thus it will be useful to select for the first equation the one that has the largest absolute value for the coefficient of x_1. After this choice is made, the elimination of x_1 from the second and third equations may be carried out. Then looking at the elimination of x_2 from the third equation, this is seen to involve the coefficient a'_{22} as a divisor. Again, to minimize round-off error in the subsequent difference operation, it would be desirable to have the largest possible magnitude for a'_{22}. As before, the second and third equations may be considered in either order, and the best choice can be made for the second equation and the elimination of x_2 from the third equation is carried out. This process is called *pivoting*.

Example of Gaussian Elimination

Consider the system of equations:

$$
\begin{array}{rrrr}
x_1 & -x_2 & -x_3 & = 2 \\
-x_1 & +2x_2 & -x_3 & = 6 \\
-x_1 & -x_2 & +3x_3 & = 8
\end{array}
$$

or, in the form of an augmented matrix of coefficients,

$$
\left[
\begin{array}{rrr|r}
1 & -1 & -1 & 2 \\
-1 & 2 & -1 & 6 \\
-1 & -1 & 3 & 8
\end{array}
\right]
$$

Because all of the coefficients of x_1 have unit magnitude, there is nothing to be gained by re-ordering the equations. For the elimination of x_1 from the second and third equations, (1) times the first equation is added to the second, and third equations:

$$
\left[
\begin{array}{rrr|r}
1 & -1 & -1 & 2 \\
0 & 1 & -2 & 8 \\
0 & -2 & 2 & 10
\end{array}
\right]
$$

At this point, it is desirable to interchange the second and third equations before eliminating x_2:

$$
\left[
\begin{array}{rrr|r}
1 & -1 & -1 & 2 \\
0 & -2 & 2 & 10 \\
0 & 1 & -2 & 8
\end{array}
\right]
$$

Adding (1/2) of the second equation to the third equation will now eliminate x_2 from the third equation:

$$\begin{bmatrix} 1 & -1 & -1 & | & 2 \\ 0 & -2 & 2 & | & 10 \\ 0 & 0 & -1 & | & 13 \end{bmatrix}$$

The unknowns are now determined in reverse order in the back substitution phase:

$$x_3 = -13$$

$$x_2 = \left(\frac{-1}{2}\right)[10 - 2(-13)] = -18$$

$$x_1 = (1)[2 + (-18) + (-13)] = -29$$

The validity of this solution is easily checked by substitution.

A subroutine to employ Gaussian elimination for the solution of a system of simultaneous linear equations is shown below. The concept of pivoting as described above is implemented in this subroutine. This particular subroutine can also be used to determine an inverse if required, depending on the various parameters in the calling statement.

The notation used in this subroutine parallels the presentation above. The matrix of coefficients is called $[A]$ within the subroutine, although it could have any name in the calling program. Further, the right side values are contained in an array $[C]$. The subroutine call is

```
call matin(A(,),C(,),sc(),n,naug,flag,DetA)
```

where flag is a control parameter with the following meanings:

flag = 0	Compute the determinant only
flag = 1	Compute the inverse of the coefficient matrix
flag = −1	Compute the solution for a system of equations

The array sc() is a work space (scratch vector) that contains no final results, but is necessary for the calculation. The right side is shown as a doubly dimensioned array b(,). The use of multiple right side vectors is discussed in more detail below. For the present, it will suffice to consider this as simply an $(n \times 1)$ right side vector. The determinant of the coefficient matrix $[A]$ is returned in the parameter DetA. The location of other results is covered in the discussion of Multiple Right Side Vectors which follows the subroutine listing.

Subroutine for Matrix Inversion
and Linear System Solution

```
sub matin(A(,),C(,),sc(),n,naug,flag,DetA)
   ! Calculation of Determinant, Matrix Inverse, and
   ! Linear System Solution by Gaussian Elimination
   ! System:              [A] [X] = [C]
```

11

```
! Augmented Matrix:    [A] = [A | C]
! n            - Number of rows
! ncol         - Number of columns in augmented matrix
! naug         - Number of right side vectors
! A(n x n)     - Coefficient array
! C(n x naug)  - Array of right side vectors
! DetA         - Value for Det[A]
! flag         - Control parameter
!       = 1    - Determine [A]^(-1) and Det[A]
!       = 0    - Determine Det[A] only
!       = -1   - Solve the linear system
! sc()         - Scratch vector, (n x 1)
nm1=n-1
ncol=n+naug

! Construct augmented matrix
! for matrix inversion
if flag=1 then
  for i=1 to n
     A(i,n+i)=1
  next i
end if

! Construct augmented matrix
! for linear system solution
if flag=-1 then
  for j=1 to naug
    for i=1 to n
       A(i,n+j)=C(i,j)
    next i
  next j
end if
algsgn=1
check=0

! Pick pivot element
for i=1 to nm1
   row=i
   Amax=abs(A(i,i))
   for k=i+1 to n
     if abs(A(k,i))>Amax then
       row=k
       Amax=abs(A(k,i))
     end if
   next k
```

```
   ! Row interchanges
   if row<>i then
      for L=i to ncol
         DetA=A(i,L)
         A(i,L)=A(row,L)
         A(row,L)=DetA
      next L
      algsgn=-algsgn
   end if
   for j=i+1 to n
      if A(j,i)<>0 then
         const=-A(j,i)/A(i,i)
         for L=i to ncol
            A(j,L)=A(j,L)+const*A(i,L)
         next L
      end if
   next j
next i

! Compute Det[A]
check=0
DetA=1
for i=1 to n
   if A(i,i)=0 then
      check=1
      exit for
   else
      DetA=DetA*A(i,i)
   end if
next i

if check=0 then DetA=algsgn*DetA
if flag=0 then exit sub
if flag=1 and check=1 then exit sub

for i=n+1 to ncol
   for k=n to 1 step -1
      sc(k)=A(k,i)
      if k<>n then
         for j=k+1 to n
            sc(k)=sc(k)-A(k,j)*sc(j)
         next j
      end if
      sc(k)=sc(k)/A(k,k)
```

```
        next k
        for L=1 to n
            A(L,i)=sc(L)
        next L
    next i
end sub
```

Multiple Right Side Vectors

There are circumstances when several solutions are required with all involving the same coefficient matrix, such as

$$[A]\{X\}_1 = \{C\}_1$$

$$[A]\{X\}_2 = \{C\}_2$$

$$[A]\{X\}_3 = \{C\}_3$$

where $\{C\}_1$, $\{C\}_2$, and $\{C\}_3$ are three distinct right side vectors and $\{X\}_1$, $\{X\}_2$, and $\{X\}_3$ are the associated solutions. If we define rectangular matrices $[X]$ and $[B]$ as

$$\underbrace{[X]}_{n \times naug} = \left[\underbrace{\{X\}_1}_{n \times 1} \mid \underbrace{\{X\}_2}_{n \times 1} \mid \underbrace{\{X\}_3}_{n \times 1} \right]_{n \times naug}$$

$$\underbrace{[C]}_{n \times naug} = \left[\underbrace{\{C\}_1}_{n \times 1} \mid \underbrace{\{C\}_2}_{n \times 1} \mid \underbrace{\{C\}_3}_{n \times 1} \right]_{n \times naug}$$

then all three systems can be expressed in one system:

$$[A][X] = \{C\}$$

with the formal solution

$$[X] = [A]^{-1}[C]$$

Considering the process of Gaussian elimination implemented via the augmented matrix of coefficients as show previously, the several right side vectors can be carried along for the elimination phase without difficulty, beginning with the augmented matrix

$$\left[\underbrace{[A]}_{n \times n} \mid \underbrace{\{C\}_1}_{n \times 1} \mid \underbrace{\{C\}_2}_{n \times 1} \mid \underbrace{\{C\}_3}_{n \times 1} \right]_{n \times (n+naug)}$$

14

Following the elimination process, the substitution is then made for each right side vector, resulting in the determination of all three solution vectors $\{X\}_1$, $\{X\}_2$, and $\{X\}_3$.

Solution Locations

The subroutine given above stores the solutions for each right side vector in the space originally occupied by the right side vector. Thus, if solutions are sought for three right side vectors as illustrated in this discussion, the augmented matrix is $[n \times (n+3)]$. The solution for the first right side vector, $\{X\}_1$, is stored in column $n+1$ of the augmented matrix, $\{X\}_2$ is stored in column $n+2$, and $\{X\}_3$ is stored in the last column, $n+3$.

Determination of an Inverse by Gaussian Elimination

Return to the statement of the single system representing several right side vectors,

$$[A]\,[X] = [C]$$

If the right side matrix, $[C]$, is in fact the identity matrix, then $[X]$ must be the inverse of $[A]$. Thus all that is required to determine an inverse is to solve the system for n right side vectors, together comprising the identity matrix. When this is done using the subroutine presented above, the inverse matrix is stored in the rightmost n columns of the augmented matrix.

System Solution Versus Inverse Matrix Determination

If the problem to be solved is the solution of a system of simultaneous linear algebraic equations, there is no doubt that the formal solution is obtained by pre-multiplying the system by the inverse of the coefficient matrix, as has been done several times earlier. On the other hand, the Gaussian elimination process does not require the explicit determination of the inverse, but rather offers the option to proceed straight to the system solution.

Which is the better way? From a purely numerical analysis perspective, it is better to go directly to the system solution, bypassing the inverse determination. This involves fewer steps and less round-off error. If fewer than n right side vectors are involved, less storage is required within the computer. This is not the end of the question, however.

Some programming languages or subroutine packages offer preprogrammed matrix operations statements, including matrix inversion. These are machine language program blocks that implement the matrix operations and are called with a single line of computer code, such as

```
mat ai=inv(a)
```

With the goal in mind of reducing human time and effort, it is often easier to use such a preprogrammed statement followed by a second statement calling for a matrix product to complete the solution, `mat x = ai*c`. This avoids the need to program and debug the subroutine and often offers faster execution time. Provided that a sufficient number of digits are carried in the calculation, typically about 14 decimal digits, the round-off problem is not significant for most matrix inversions. Thus it is a trade-off that must be made, usually in favor of speed and economy of human effort, but with a careful eye to the validity of the results.

When does a linear system have a solution?

As we work with linear systems, we find that they often have solutions, but there are many that do not have solutions. This can be pretty disconcerting, and we need to have ways to know when to expect to be able to find a solution. We will address this question here by means of some simple examples.

Consider the linear system

$$\begin{bmatrix} a_{11} & a_{12} \\ a_{21} & a_{22} \end{bmatrix} \begin{Bmatrix} x_1 \\ x_2 \end{Bmatrix} = \begin{Bmatrix} c_1 \\ c_2 \end{Bmatrix}$$

This can be written without matrix notation as the two equations

$$a_{11}x_1 + a_{12}x_2 = c_1$$
$$a_{21}x_1 + a_{22}x_2 = c_2$$

Each of these equations represents a straight line in the x_1–x_2 plane. If c_1 and c_2 are both zero, the system is said to be *homogenous*. If c_1 and c_2 are not both zero, then the system is called *nonhomogenous*.

What are the possibilities for two straight lines in the plane? There are two cases:

Case 1: The two lines may intersect at one point, giving a unique solution, or

Case 2: The two lines may be parallel, in which case we expect no solution.

We need to look more carefully at each of these cases.

Case 1: Intersecting Lines

If the two lines intersect, their slopes must be unequal. To check for this, put both equations into the slope-intercept form, treating x_2 as dependent on x_1:

$$x_2 = -\frac{a_{11}}{a_{12}}x_1 + \frac{c_1}{a_{12}}$$
$$x_2 = -\frac{a_{21}}{a_{22}}x_1 + \frac{c_2}{a_{22}}$$

The condition for intersection is that the two slopes be unequal,

$$-\frac{a_{11}}{a_{12}} \neq -\frac{a_{21}}{a_{22}}$$

or, after some rearrangement,

$$a_{11}a_{22} - a_{12}a_{21} \neq 0$$

The expression on the left side of the inequality is readily recognized as the determinant of the $[A]$ matrix. In this case, and by extension in all cases, the condition

16

for an intersection is simply that the determinant of the coefficient matrix must be non-zero.

Note also that if the determinant is almost zero, the two lines have nearly equal slopes. While there is still exactly one point of intersection, in a numerical solution carrying only a limited number of decimal places, it may be very difficult to locate the actual solution.

The development above has assumed implicitly that the system of equation is nonhomogenous. In the event that the system is actually homogenous, then each line represents a straight line through the origin. Since two straight lines can intersect in only one point at most, *the origin is the one and only solution point.* This is often called the *trivial solution.* This is because we are usually seeking a solution with one or more of the unknowns having a non-zero value.

Case 2: Parallel Lines

Now consider the case where the two lines are parallel. From the analysis done in Case 1, *the condition for parallelism is a zero determinant for the coefficient matrix.* Can there still be a solution? It depends on the values of the two intercepts, c_1/a_{12} and c_2/a_{22}.

- If the intercepts are different, then the two lines are simply parallel and have no points in common.

- If, in addition to being parallel, the two intercepts are the same, then the two equations actually represent the same straight line and have all points in common. (They are said to be *linearly dependent.*) This means that there are infinitely many pairs of points (x_1, x_2) that satisfy both equations, and a unique solution is not available.

If the system of equations is homogenous, then the intercepts are each zero. This means that $(0,0)$ is definitely a solution (the trivial solution). However, since the two equations actually represent the same straight line, all points on that line are solutions, and $(0,0)$ is only one of them. There are infinitely many nontrivial solutions. Thus the important theorem (without proof),

> *A nontrivial solution exists for a system of homogenous, linear equations if, and only if, the determinant of the coefficient matrix is zero.*

Note that this theorem does not give directions for finding the non-trivial solutions; it only states the conditions under which they exist. To find them, we may assign a value to one of the variables, say x_1, and then use either equation to compute the value of x_2. If we assign $x_1 = n_1$ and from one of the equations determine that the associated value of $x_2 = n_2$ then (n_1, n_2) is *a solution.* Note, however, that for any number, k, the pair (kn_1, kn_2) is also a solution.

References

Conte, S.D., *Elementary Numerical Analysis,* McGraw-Hill, 1965.

Frazer, R.A., Duncan, W.J., and Collar, A.R., *Elementary Matrices,* Cambridge Press, 1965.

Hohn, F.E., *Elementary Matrix Algebra,* Macmillan, 1964.

Steen, F.H, and Ballou, D.H., *Analytic Geometry,* Ginn & Co., 1955.

Chapter 3

Vectors

Vectors in Two and Three Dimensions, Notations

Vectors are usually described as directed line segments, often visualized as an arrow beginning at the origin of coordinates and going to a head end point. Every vector has two essential characteristics: (1) magnitude and (2) direction. The length of the line segment is the magnitude of the vector, and the direction of the line segment is the direction of the vector. This is in contrast to the familiar quantity called a *scalar* that has a magnitude (possibly including an algebraic sign) but no direction.

One of the useful concepts used in expressing vectors is that of the *unit vector*. As the name suggests, this is a vector of unit length but having a particular direction. The basic function of a unit vector is to carry direction information while having unit length. There are many commonly defined unit vectors, that is, unit vectors that have (almost) universally agreed definitions. The user should understand that new unit vectors may be defined at will, whenever it suits the user's purpose. It is important, however, to make clear how all unit vectors are defined.

Among the widely understood unit vectors are three unit vectors lying along the coordinate axes for a rectangular Cartesian coordinate system. These are usually denoted as $\mathbf{i}$, $\mathbf{j}$, and $\mathbf{k}$, where

$$\mathbf{i} = \text{unit vector along the positive } x\text{-axis}$$
$$\mathbf{j} = \text{unit vector along the positive } y\text{-axis}$$
$$\mathbf{k} = \text{unit vector along the positive } z\text{-axis}$$

Consider a vector, $\mathbf{V}$, in three dimensions that has its foot at the origin and its head at the point (a, b, c), where a, b, c are coordinates referred to a rectangular, Cartesian coordinate system. Using the idea of unit vectors, this vector can be written as

$$\mathbf{V} = a\mathbf{i} + b\mathbf{j} + c\mathbf{k}$$
$$\vec{V} = a\vec{i} + b\vec{j} + c\vec{k}$$

where these two statements are exactly equivalent. Bold face type is commonly used for vectors in printed materials, but the use of an arrow or bar over the letter is the common way to denote a vector in handwritten work. If only two dimensions are being considered, there will be only two coefficients, a and b, and two unit vectors, **i** and **j**. The vectors **i**, **j**, and **k** are often called *Cartesian base vectors*.

In the expression above for the vector **V** above, the coefficients a, b, and c are scalars that multiply the unit vectors. In the first product, the result is a vector in the same direction as the unit vector, but a times the length of **i**, which is simply $a \times 1$ or a in length. The products $b\mathbf{j}$ and $c\mathbf{k}$ produce vectors in the directions of **j** and **k**, respectively. Each of the three terms in the expression for **V** is called a component of **V**, so that $a\mathbf{i}$ is the component along the x-axis, $b\mathbf{j}$ is the component along the y-axis, and $c\mathbf{k}$ is the component along the z-axis. Note that each of these components is itself a vector, each directed along one of the coordinate axes.

In the case of the vector components described above, the unit vectors are not particularly useful, as long as we have some other means to denote which component lies along which axis. All of the length information, including sign, is carried in the coefficients, a, b, and c. The tracking of which coefficient is associated with which direction can be done more conveniently by place, that is, by the location assigned to a particular component within an array. Thus, consider the matrix representation of **V** as a column of three components:

$$\{V\} = \left\{ \begin{array}{c} a \\ b \\ c \end{array} \right\}$$

The representation of a vector as an arrow whose length is the magnitude of the vector and whose direction is that of the vector is useful for representing the vector geometrically. In order to move on to analytical geometry the representation in terms of the unit vectors is useful. Because computers are very well suited to handling arrays, the matrix representation is the preferred way to handle a vector within the computer. Computers make no distinction between row and column vectors, but it will be useful to think of all vectors as being represented as column vectors within the computer.

Vector Arithmetic Example

Consider the following two vectors in two dimensions:

A, a vector of magnitude 5 at 45° above the x-axis, and

B, a vector of magnitude 3 at 120° CCW (counterclockwise) from the positive x-axis.

The following operations are required:

1. Write each of these vectors in terms of numerical coefficients and the unit vectors **i**, **j**, and **k**

2. Determine the vector components of the sum $\mathbf{A} + \mathbf{B}$ and the difference $\mathbf{B} - \mathbf{A}$

3. Write each of the original vectors in matrix form

4. Determine, in matrix form, the vector sum $\mathbf{A} + \mathbf{B}$ and the difference $\mathbf{B} - \mathbf{A}$.

Solution:

Item 1

$$
\begin{aligned}
\mathbf{A} &= 5 \cdot \cos(45^\circ)\mathbf{i} + 5 \cdot \sin(45^\circ)\mathbf{j} = 3.5355\mathbf{i} + 3.5355\mathbf{j} \\
\mathbf{B} &= 3 \cdot \cos(120^\circ)\mathbf{i} + 3 \cdot \sin(120^\circ)\mathbf{j} = -1.5000\mathbf{i} + 2.5981\mathbf{j}
\end{aligned}
$$

Item 2

$$
\begin{aligned}
\mathbf{A} + \mathbf{B} &= (3.5355\mathbf{i} + 3.5355\mathbf{j}) + (-1.5000\mathbf{i} + 2.5981\mathbf{j}) \\
&= [3.5355 + (-1.5000)]\mathbf{i} + (3.5355 + 2.5981)\mathbf{j} \\
&= 2.0355\mathbf{i} + 6.1336\mathbf{j}
\end{aligned}
$$

$$
\begin{aligned}
\mathbf{B} - \mathbf{A} &= (-1.5000\mathbf{i} + 2.5981\mathbf{j}) - (3.5355\mathbf{i} + 3.5355\mathbf{j}) \\
&= [(-1.5000) - (3.5355)]\mathbf{i} + [2.5981 - (3.5355)]\mathbf{j} \\
&= -5.0355\mathbf{i} - 0.9374\mathbf{j}
\end{aligned}
$$

Item 3

$$
\begin{aligned}
\{A\} &= \left\{ \begin{array}{c} 5 \cdot \cos(45^\circ) \\ 5 \cdot \sin(45^\circ) \end{array} \right\} = \left\{ \begin{array}{c} 3.5355 \\ 3.5355 \end{array} \right\} \\
\{B\} &= \left\{ \begin{array}{c} 3 \cdot \cos(120^\circ) \\ 3 \cdot \sin(120^\circ) \end{array} \right\} = \left\{ \begin{array}{c} -1.5000 \\ 2.5981 \end{array} \right\}
\end{aligned}
$$

Item 4

$$
\begin{aligned}
\mathbf{A} + \mathbf{B} &= \{A\} + \{B\} \\
&= \left\{ \begin{array}{c} 3.5355 \\ 3.5355 \end{array} \right\} + \left\{ \begin{array}{c} -1.5000 \\ 2.5981 \end{array} \right\} \\
&= \left\{ \begin{array}{c} 2.0355 \\ 6.1336 \end{array} \right\} \\
\mathbf{B} - \mathbf{A} &= \{B\} - \{A\} \\
&= \left\{ \begin{array}{c} -1.5000 \\ 2.5981 \end{array} \right\} - \left\{ \begin{array}{c} 3.5355 \\ 3.5355 \end{array} \right\} \\
&= \left\{ \begin{array}{c} -5.0355 \\ -0.9374 \end{array} \right\}
\end{aligned}
$$

The Dot Product of Vectors

There are several sides to the multiplication of vectors. One of these has already been employed without comment: the multiplication of a vector by a scalar. This was done in connection with unit vectors used to express a vector, $\mathbf{V}$, where products such as $a\mathbf{i}, b\mathbf{j}$, and $c\mathbf{k}$ appeared. The coefficients a, b, and c are scalars, while $\mathbf{i}, \mathbf{j}$, and $\mathbf{k}$ are each unit vectors. The meaning of this sort of multiplication of vectors is clear: the scalar simply multiplies the magnitude of the vector to produce a new vector having the same direction as the unit vector and a magnitude that is equal to the scalar multiplied by the magnitude of the unit vector. There are two other types of multiplication involving vectors: the *dot product* to be discussed here and the *cross (or vector) product* to be discussed later.

The dot product is a product of two vectors that produces a scalar result. This product is denoted by a dot between the two vector factors, and hence its name. The scalar result is defined to be the product of three factors: (1) the magnitude of the first vector factor, (2) the magnitude of the second vector factor, and (3) the cosine of the angle between the vectors, thus:

$$\mathbf{V}_1 \cdot \mathbf{V}_2 = |\mathbf{V}_1|\,|\mathbf{V}_2|\cos\left(\mathbf{V}_1, \mathbf{V}_2\right)$$

where the absolute value bars, $|\ \ |$ around the two vector factors denote the magnitude of the vector and $(\mathbf{V}_1, \mathbf{V}_2)$ denotes the angle from vector $\mathbf{V}_1$ to vector $\mathbf{V}_2$. Consider this definition applied to two vectors expressed in terms of the unit vectors $\mathbf{i}, \mathbf{j}$, and $\mathbf{k}$:

$$A = a_1\mathbf{i} + a_2\mathbf{j} + a_3\mathbf{k}$$
$$B = b_1\mathbf{i} + b_2\mathbf{j} + b_3\mathbf{k}$$

$$\mathbf{A} \cdot \mathbf{B} = (a_1\mathbf{i} + a_2\mathbf{j} + a_3\mathbf{k}) \cdot (b_1\mathbf{i} + b_2\mathbf{j} + b_3\mathbf{k})$$

$$\begin{aligned}= \ &a_1\mathbf{i} \cdot (b_1\mathbf{i} + b_2\mathbf{j} + b_3\mathbf{k})\\ &+a_2\mathbf{j} \cdot (b_1\mathbf{i} + b_2\mathbf{j} + b_3\mathbf{k})\\ &+a_3\mathbf{k} \cdot (b_1\mathbf{i} + b_2\mathbf{j} + b_3\mathbf{k})\end{aligned}$$

$$\begin{aligned}= \ &a_1\mathbf{i} \cdot b_1\mathbf{i} + a_1\mathbf{i} \cdot b_2\mathbf{j} + a_1\mathbf{i} \cdot b_3\mathbf{k}\\ &+a_2\mathbf{j} \cdot b_1\mathbf{i} + a_2\mathbf{j} \cdot b_2\mathbf{j} + a_2\mathbf{j} \cdot b_3\mathbf{k}\\ &+a_3\mathbf{k} \cdot b_1\mathbf{i} + a_3\mathbf{k} \cdot b_2\mathbf{j} + a_3\mathbf{k} \cdot b_3\mathbf{k}\end{aligned}$$

Look at the mixed dot product

$$\mathbf{i} \cdot \mathbf{j} = (1)(1)\cos\left(90^o\right) = (1)(1)(0) = 0$$

The same zero value will result from all of the mixed dot products, so terms involving those products will vanish from the dot product $\mathbf{A} \cdot \mathbf{B}$. Similarly, look at the product of a unit vector with itself, such as

$$\mathbf{i} \cdot \mathbf{i} = (1)(1)\cos\left(0^o\right) = (1)(1)(1) = 1$$

Similar results are obtained for the products of the other two unit vectors with themselves. Thus, it is evident that the dot product of each of the base vectors with itself is one. Returning to the dot product $\mathbf{A} \cdot \mathbf{B}$, the result is reduced to

$$\mathbf{A} \cdot \mathbf{B} = a_1 b_1 + a_2 b_2 + a_3 b_3$$

Notice that the form of the result *is a sum of products of the scalar coefficients of corresponding vector components.* This is exactly the sum produced by the matrix product $\{A\}^t \{B\}$. In order to perform a vector dot product within the computer, all that is required is the matrix product of the transpose of the first factor with the second factor.

One important application of the dot product is in finding the magnitude of a vector for which the components are known. For $\mathbf{A}$ as above, the dot product of $\mathbf{A}$ with itself is

$$\mathbf{A} \cdot \mathbf{A} = a_1^2 + a_2^2 + a_3^2$$

which is recognized as the square of the magnitude of the vector, according to the Pythagorean theorem extended to three dimensions. This leads to the very useful result

$$Mag\,(\mathbf{A}) = |\mathbf{A}| = \sqrt{\mathbf{A} \cdot \mathbf{A}}$$

Dot Product Example Calculations

Consider the two vectors $\mathbf{P}$ and $\mathbf{Q}$:

$$\mathbf{P} = \mathbf{i} + 2\mathbf{k}$$
$$\mathbf{Q} = \mathbf{i} + \mathbf{j} - 3\mathbf{k}$$

Determine (1) the magnitude of the vector $\mathbf{Q}$ and (2) the angle between $\mathbf{P}$ and $\mathbf{Q}$. For the magnitude of $\mathbf{Q}$,

$$
\begin{aligned}
|\mathbf{Q}|^2 &= Q_x^2 + Q_y^2 + Q_z^2 \\
&= 1^2 + 1^2 + (-3)^2 \\
&= 11 \\
|\mathbf{Q}| &= \sqrt{11}
\end{aligned}
$$

For the angle between the two vectors, first consider the dot product definition:

$$\mathbf{P} \cdot \mathbf{Q} = |\mathbf{P}|\,|\mathbf{Q}|\cos{(\mathbf{P}, \mathbf{Q})}$$

from which

$$\cos{(\mathbf{P}, \mathbf{Q})} = \frac{\mathbf{P} \cdot \mathbf{Q}}{|\mathbf{P}||\mathbf{Q}|}$$

$$= \frac{(\mathbf{i} + 2\mathbf{k}) \cdot (\mathbf{i} + \mathbf{j} - 3\mathbf{k})}{\sqrt{1^2 + 2^2} \cdot \sqrt{11}}$$

$$= \frac{1\mathbf{i} \cdot \mathbf{i} - 6\mathbf{k} \cdot \mathbf{k}}{\sqrt{55}}$$

$$= \frac{-5}{\sqrt{55}} \approx -0.6742$$

$$\angle{(\mathbf{P}, \mathbf{Q})} \approx 132.39^{o}$$

The Cross Product of Two Vectors

The final type of vector multiplication to be considered is the cross product, also called the vector product. Both names are descriptive in that the notation used for this product is a $\times$, and the result of this product is another vector. The magnitude of the result is determined by

$$|\mathbf{A} \times \mathbf{B}| = |\mathbf{A}| \, |\mathbf{B}| \sin{(\mathbf{A}, \mathbf{B})}$$

Note that this is similar to the expression for the dot product, but that the sine of the angle between the two vectors is required, rather than the cosine used in the dot product expression. The direction of the resulting vector is determined by the right hand rule:

If the fingers of the right hand are first aligned with the first vector factor and then curled toward the second vector, the thumb of the right hand points in the direction of the cross product result.

While this completely describes the cross product on one level, it does not lend itself to use in analysis, and even less to computer calculation. For vectors expressed in terms of rectangular Cartesian coordinates, there is a convenient way to express the cross product as a determinant. For two vectors $\mathbf{A}$ and $\mathbf{B}$

$$\mathbf{A} = a_1\mathbf{i} + a_2\mathbf{j} + a_3\mathbf{k}$$
$$\mathbf{B} = b_1\mathbf{i} + b_2\mathbf{j} + b_3\mathbf{k}$$

the cross product is expressed as

$$\mathbf{A} \times \mathbf{B} = \begin{vmatrix} \mathbf{i} & \mathbf{j} & \mathbf{k} \\ a_1 & a_2 & a_3 \\ b_1 & b_2 & b_3 \end{vmatrix} = \begin{array}{l} \mathbf{i}\,(a_2 b_3 - a_3 b_2) \\ +\mathbf{j}\,(a_3 b_1 - a_1 b_3) \\ +\mathbf{k}\,(a_1 b_2 - a_2 b_1) \end{array}$$

Note the manner in which the determinant is constructed. The first row consist of the base vectors, $\mathbf{i}$, $\mathbf{j}$, and $\mathbf{k}$. The components of the first vector make up the second row, and the components of the second vector occupy the third row. After the determinant is assembled, the conventional means are employed to evaluate the determinant by multiplying along the diagonals.

Cross Product Example Calculations

Consider the two vectors $\mathbf{P}$ and $\mathbf{Q}$

$$\mathbf{P} = \mathbf{i} + 2\mathbf{k}$$
$$\mathbf{Q} = \mathbf{i} + \mathbf{j} - 3\mathbf{k}$$

Use the cross product to (1) determine the angle between the vectors $\mathbf{P}$ and $\mathbf{Q}$, and (2) determine a unit vector normal to the plane defined by the two vectors.

From the previous example involving dot products, the magnitudes of these two vectors are known to be $\sqrt{5}$ and $\sqrt{11}$, respectively. Thus, unit vectors in the directions of $\mathbf{P}$ and $\mathbf{Q}$ may be written as

$$\mathbf{u}_P = \left(\frac{1}{\sqrt{5}}\right)\mathbf{P} = \frac{\mathbf{i} + 2\mathbf{k}}{\sqrt{5}}$$
$$\mathbf{u}_Q = \left(\frac{1}{\sqrt{11}}\right)\mathbf{Q} = \frac{\mathbf{i} + \mathbf{j} - 3\mathbf{k}}{\sqrt{11}}$$

Now look at the magnitude of the cross product of these two vectors:

$$|\mathbf{u}_P \times \mathbf{u}_Q| = |\mathbf{u}_P||\mathbf{u}_Q|\sin(\mathbf{P}, \mathbf{Q})$$

$$= (1)(1)\sin(\mathbf{P}, \mathbf{Q})$$

$$= \sin(\mathbf{P}, \mathbf{Q})$$

$$\mathbf{u}_P \times \mathbf{u}_Q = \begin{vmatrix} \mathbf{i} & \mathbf{j} & \mathbf{k} \\ \frac{1}{\sqrt{5}} & 0 & \frac{2}{\sqrt{5}} \\ \frac{1}{\sqrt{11}} & \frac{1}{\sqrt{11}} & \frac{-3}{\sqrt{11}} \end{vmatrix} = -\frac{2}{\sqrt{55}}\mathbf{i} + \frac{5}{\sqrt{55}}\mathbf{j} + \frac{1}{\sqrt{55}}\mathbf{k}$$

The magnitude of this vector is determined using a dot product:

$$|\mathbf{u}_P \times \mathbf{u}_Q| = \frac{1}{\sqrt{55}}\sqrt{(-2)^2 + (+5)^2 + (1)^2} = \sqrt{\frac{30}{55}} \approx 0.7385$$

Since this value is also the sine of the angle between the vectors, the angle can be determined using an arcsine function, bearing in mind that the desired result may not be the principal value of the arcsine:

$$\angle(\mathbf{P}, \mathbf{Q}) \approx \arcsin(0.7385)$$
$$= 180° - \text{Arcsin}(0.7385)$$
$$= 180° - 47.61°$$
$$= 132.39°$$

The cross product $\mathbf{u}_P \times \mathbf{u}_Q$ defined a vector normal to the plane of $\mathbf{P}$ and $\mathbf{Q}$, but it was not a unit vector. Looking back over the work above, the magnitude of this vector is the sine of the angle between the two vectors. A unit normal vector can be determined by simply normalizing (adjusting to unit magnitude) the cross product:

$$
\begin{aligned}
\mathbf{u_{\perp P \perp Q}} &= \frac{\mathbf{u}_P \times \mathbf{u}_Q}{|\mathbf{u}_P \times \mathbf{u}_Q|} = \frac{\mathbf{u}_P \times \mathbf{u}_Q}{\sqrt{\frac{30}{55}}} \\
&= \sqrt{\frac{55}{30}} \left(-\frac{2}{\sqrt{55}}\mathbf{i} + \frac{5}{\sqrt{55}}\mathbf{j} + \frac{1}{\sqrt{55}}\mathbf{k} \right) \\
&= -\frac{2}{\sqrt{30}}\mathbf{i} + \frac{5}{\sqrt{30}}\mathbf{j} + \frac{1}{\sqrt{30}}\mathbf{k}
\end{aligned}
$$

That this is in fact a unit vector may be easily verified by summing the squares of the components.

It is widely accepted that there is no matrix equivalent to a cross product, but that is a misconception. In the area of tensor analysis, of which matrix analysis is subset, the cross product is expressed using the *permutation symbol*, usually denoted as e_{ijk}, which suggest the need for a three-dimensional array. The permutation symbol takes three different values, according to the values of the indices:

$$
\begin{aligned}
e_{ijk} &= +1 \quad \text{for } ijk = 123, 231, 312 \\
e_{ijk} &= -1 \quad \text{for } ijk = 132, 213, 321 \\
e_{ijk} &= 0 \quad\;\; \text{for any repeated subscripts} \\
&\qquad\quad\;\; \text{such as } 112, 122, 133, \ldots
\end{aligned}
$$

Using the permutation symbol, the cross product is expressed as

$$
\mathbf{A} \times \mathbf{B} = \sum_{i=1}^{i=3} \sum_{j=1}^{j=3} e_{ijk} a_j b_k
$$

Most computer languages can accept arrays with three indices, so that implementation of this expression is a possible way to compute the cross product of two vectors. There is, however, an easier way that is described below.

Returning to the expanded determinant for the cross product, consider the process of writing that result as the product of a square matrix with a column vector, thus:

$$
\begin{aligned}
\mathbf{A} \times \mathbf{B} &= \begin{aligned} &\mathbf{i}\,(a_2 b_3 - a_3 b_2) \\ &+\mathbf{j}\,(a_3 b_1 - a_1 b_3) \\ &+\mathbf{k}\,(a_1 b_2 - a_2 b_1) \end{aligned} \\
&= \underbrace{\begin{bmatrix} 0 & -a_3 & a_2 \\ a_3 & 0 & -a_1 \\ -a_2 & a_1 & 0 \end{bmatrix}}_{[A\times]} \begin{Bmatrix} b_1 \\ b_2 \\ b_3 \end{Bmatrix} \\
&= [A\times]\{B\}
\end{aligned}
$$

26

The (3×3) matrix $[A\times]$ is simply defined to be whatever is required to make the matrix-vector product equal to the original cross product. The elements of $[A\times]$ are determined entirely by inspection of the expanded determinant and are found to be simply the elements of $\{A\}$ placed in various off-diagonal locations and in some cases with sign changes. There is no generally accepted matrix operation to determine $[A\times]$ from $\{A\}$, but a few simple lines of computer code will do what is necessary.

Note that all of the diagonal elements of the matrix $[A\times]$ are zeroes, and further, corresponding off-diagonal elements have opposite signs but the same magnitudes. These are the marks of a *skew symmetric matrix*, and $[A\times]$ is definitely skew symmetric. In contrast, for a symmetric matrix, the diagonal elements may have any values and the corresponding off-diagonal elements will be equal in each case.

For computer implementation, this approach requires only that the elements of the first vector factor be assigned to the proper locations in the matrix $[A\times]$, and then that a matrix product be carried out to form the product $[A\times]\{B\}$. A subroutine to make the required assignments is given below.

Subroutine to Develop the Matrix $[A\times]$

```
sub ACross(A(),Ax(,))
    ! A subroutine to develop
    ! the (3 x 3) matrix [Ax]
    ! from the column vector {A}
    ! Initialize the array to zeroes
    mat Ax=zer
    ! Assign elements of [Ax]
    Ax(1,2)=-A(3)
    Ax(1,3)=A(2)
    Ax(2,1)=A(3)
    Ax(2,3)=-A(1)
    Ax(3,1)=-A(2)
    Ax(3,2)=A(1)
end sub
```

Example of Dot and Cross Products Done by Computer

The determination of the angles between the vectors **P** and **Q** was done by hand calculation above. The following computer code is typical of that required to perform these calculations by means of the computer, employing the matrix product subroutine shown and the subroutine for $[A\times]$ given above.

Examples of Computer Dot and Cross Product Calculations

```
!                DotCross.Tru
option nolet
option base 1
dim p(3),q(3),up(3),uq(3),b(3),px(3,3)
clear
```

```
p(1)=1
p(3)=2
q(1)=1
q(2)=1
q(3)=-3

! Determine |P| and |Q|
call DotProd(p(),p(),psq)
magP=sqr(psq)
call DotProd(q(),q(),qsq)
magQ=sqr(qsq)

! Scale P & Q to form unit vectors
for i=1 to 3
    up(i)=p(i)/magP
    uq(i)=q(i)/magQ
next i

! Dot Product
call DotProd(up(),uq(),cospq)
angpq1=acos(cospq)
lbl1$="From dot product, angle P,Q = "
lbl2$="                                = "
print lbl1$;angpq1;" rad"
print lbl2$;angpq1*180/pi;" deg"
print

! Cross Product
! Matrix product of px with uq
call ACross(up(),px(,))
mat b=px*uq
call DotProd(b(),b(),bsq)
magB=sqr(bsq)
angpq2=asin(magB)
lbl3$="From cross product, angle P,Q = "
lbl4$="                                = "
print lbl3$;angpq2;" rad"
print lbl4$;angpq2*180/pi;" deg"
end

sub DotProd(a(),b(),c)
    ! Dot product of two vectors
    sum=0
    for i=1 to 3
        sum=sum+a(i)*b(i)
```

```
      next i
      c=sum
end sub

sub ACross(A(),Ax(,))
    ! A subroutine to develop
    ! the (3 x 3) matrix [Ax]
    ! from the column vector {A}
    ! Initialize the array to zeroes
    mat Ax=zer
    ! Assign elements of [Ax]
    Ax(1,2)=-A(3)
    Ax(1,3)=A(2)
    Ax(2,1)=A(3)
    Ax(2,3)=-A(1)
    Ax(3,1)=-A(2)
    Ax(3,2)=A(1)
end sub
```

The results of this sample program are:
```
From dot product, angle P,Q = 2.3106771  rad
                             = 132.39205  deg

From cross product, angle P,Q = .83091555  rad
                               = 47.607954  deg
```

The reason for the apparent error in the angle when computed by means of the cross product lies in the arcsine function. Computer codes generally produce the principal value of the inverse trigonometric functions, in this case the principal value of the arcsine. What is needed is not the principal value, but rather the value of the arcsine falling in the second quadrant. Quite a bit more program logic would have to be provided to recognize this fact and make the proper adjustment.

At this point, the usefulness of matrix notation for implementing vector calculations within the computer should be quite apparent. It is clearly quite valuable to be able to move freely between the geometric description of a vector (magnitude and direction), the analytical description of that same vector (components resolved on base vectors associated with a coordinate system), and matrix notation (vectors considered as column arrays).

References

Lass, H., *Vector and Tensor Analysis*, McGraw-Hill, 1950.

Chapter 4

Curve Fitting

The term *curve fitting* is rather descriptive and refers to the process of determining an equation that will represent a particular set of data points. The *curve* under discussion is the graphical form resulting when the fitted equation is plotted. One might well ask why anyone would want to do this, and there are several answers that may be given:

- Smoothing. Measured data is often rather erratic, and smoothness is expected (rightly or wrongly!) in many physical phenomena; therefore, the process of fitting a curve may be done for the purpose of smoothing. This application will be discussed in more detail in Chapter 5.

- Removal of Ambiguity. Measurements are often repeated giving rise to multiple data values, even when we expect only a single curve to describe the phenomenon. This happens, for example, when a stress-strain curve is measured, recording data first while increasing the load on the test specimen and secondly during the unloading process. The loading and unloading values rarely coincide (in fact, they cannot due to hysteresis), but a single valued stress-strain curve is to be determined, presumably falling between the loading and unloading curves. The desired single valued curve is determined by some sort of curve fitting technique.

- Mathematical Summary. Measured data is unwieldy for use in mathematical analysis. It is often desirable to have a mathematical description that represents, in some sense, a summary of the measured results and from which approximate values can be obtained for any point in the domain of the original data. A curve fit provides the required mathematical description for use elsewhere.

- Identification of Parameters. Mathematical models often include constants that can only be evaluated by experimental measurements. The process of fitting the mathematical model to the measured data provides the values for the previously unknown constants. The constants are said to be "identified" in the fitting process.

In the freshman physics labs of years gone by, the instructions for data reduction often said something like, "Determine a way to plot your data as a straight line." On first encounter, this often seemed like a matter beyond the control of the student. The data, a set of number pairs $\{x_i, y_i\}$, either lay along a straight line or they did not. The intent of the instruction, however, was that the student would explore possible plotting alternatives, such as plotting y^2 versus x, $\ln y$ versus x, etc. If a suitable form could be determined, then a relation could be easily drawn from the experimental data, perhaps a relation that would not have been easily determined by simply looking at the plot of y versus x. Acton (pp. 25 – 28) describes this process in his book on numerical methods in some detail and points out that this process enables quite complicated relations to be represented by relatively simple curves. There is much to be said for doing this sort of "adjustment," before any mathematical curve fitting is begun.

When any initial "adjustment" of the data has been completed, there is still the question of how to do the mathematical curve fitting. The most widely used method for curve fitting is called *least squares* curve fitting. It is a very basic tool and yields good results when intelligently applied. It will be the principal thrust of this chapter.

Basic Concept

The term *least squares* is only partially descriptive, and it is important to get a clear understanding of just what is to be done. Consider the measured data to be a set of number pairs $\{x_i, y_i\}$, $i = 1, 2, \ldots N_p$. The fitted curve will be some function, $f(x)$, perhaps suggested by theoretical considerations or often simply by the shape of a plot of y_i versus x_i. The curve will not be required to pass exactly through any single data point, much less through all of them. It must, however, pass "close" to all of the data, in some sense. At any particular data point, say (x_i, y_i), there is usually a difference between value defined by the fitted curve and the data point. This is often called an "error," and it is defined by

$$\varepsilon_i = f(x_i) - y_i$$

If we think in terms of a simple Cartesian plot of the data, with x-values plotted horizontally and y-values plotted vertically, then ε_i is the difference between $f(x_i)$ and the data value y_i measured in the vertical direction. If the function f is to pass "close" to all of the data, it will be necessary to avoid large values for any of the ε_i. There may appear to be a number of possible ways to accomplish this, but the way that has achieved universal acceptance is to minimize, in a mathematical sense, the sum of the squares of the errors. Thus we can define a figure of merit (FOM) that represents the quality of the fit:

$$FOM = \sum_{i=1}^{N_p} \varepsilon_i^2 = \sum_{i=1}^{N_p} [f(x_i) - y_i]^2$$

31

The figure of merit cannot be negative by virtue of the manner in which it is defined, but the smaller it is, the better is the quality of the fit that has been achieved. But how do we minimize? What is available to be tweaked?

The function f must contain some undetermined parameters, say c_j, that can be adjusted to accomplish the minimization. Thus the equations describing the minimum condition are

$$\frac{\partial (FOM)}{\partial c_j} = 2 \sum_{i=1}^{N_p} [f(x_i) - y_i] \frac{\partial f}{\partial c_j} = 0 \qquad j = 1, 2, \ldots N_c$$

where N_c is the number of parameters available to be varied. There will be N_c equations, to be solved for the values of the c's. Thus, the term *least squares* refers to a process of curve fitting for which a figure of merit is defined as the sum of the squares of the errors between the data and the fitted curve, and that figure of merit is minimized with respect to the available parameters of the fitted function.

Polynomial Least Squares

One of the most common types of least square fitting employs a polynomial for the fitted function, and the polynomial coefficients are the parameters available for minimization. Thus,

$$f(x) = c_0 + c_1 x + c_2 x^2 + \ldots + c_N x^N$$

where $N = N_c - 1$. For this function, the error is

$$\varepsilon_i = f(x_i) - y_i = c_0 + c_1 x + c_2 x^2 + \ldots + c_N x^N - y_i$$

and the figure of merit is

$$FOM = \sum_{i=1}^{N_p} \left[c_0 + c_1 x + c_2 x^2 + \ldots + c_N x^N - y_i \right]^2$$

In order to minimize the figure of merit, it is differentiated with respect to each of the coefficient, and that derivative set equal to zero, all as indicated above. The result is

$$\frac{\partial (FOM)}{\partial c_0} = \sum_{i=1}^{N_p} \left[c_0 + c_1 x_i + c_2 x_i^2 + \ldots + c_N x_i^N - y_i \right] (1) = 0$$

$$\frac{\partial (FOM)}{\partial c_1} = \sum_{i=1}^{N_p} \left[c_0 + c_1 x_i + c_2 x_i^2 + \ldots + c_N x_i^N - y_i \right] (x_i) = 0$$

$$\frac{\partial (FOM)}{\partial c_2} = \sum_{i=1}^{N_p} \left[c_0 + c_1 x_i + c_2 x_i^2 + \ldots + c_N x_i^N - y_i \right] (x_i^2) = 0$$

$$\vdots$$

After some re-arrangement, this system of equations can be written as

$$
\begin{aligned}
c_0 \sum (1) + c_1 \sum x_i + c_2 \sum x_i^2 + \ldots + c_N \sum x_i^N &= \sum y_i \\
c_0 \sum x_i + c_1 \sum x_i^2 + c_2 \sum x_i^3 + \ldots + c_N \sum x_i^{N+1} &= \sum x_i y_i \\
c_0 \sum x_{i^2} + c_1 \sum x_i^3 + c_2 \sum x_i^4 + \ldots + c_N \sum x_i^{N+2} &= \sum x_i^2 y_i \\
&\vdots
\end{aligned}
$$

or, in matrix form,

$$
\begin{bmatrix}
N & \sum x_i & \sum x_i^2 & \cdots & \sum x_i^N \\
\sum x_i & \sum x_i^2 & \sum x_i^3 & \cdots & \sum x_i^{N+1} \\
\sum x_i^2 & \sum x_i^3 & \sum x_i^4 & \cdots & \sum x_i^{N+2} \\
\vdots & \vdots & \vdots & & \vdots
\end{bmatrix}
\begin{Bmatrix}
c_0 \\ c_1 \\ c_2 \\ \vdots \\ c_N
\end{Bmatrix}
=
\begin{Bmatrix}
\sum y_i \\ \sum x_i y_i \\ \sum x_i^2 y_i \\ \vdots
\end{Bmatrix}
$$

This system of equations is linear and readily solvable by the methods discussed in Chapter 1. Note that in computing the $(N+1) \times (N+1)$ coefficient matrix, it is only necessary to evaluate $2N+1$ sums (rather than $(N+1)^2 = N^2 + 2N + 1$ sums), because only this many are unique. It is generally suggested that the unique sums be evaluated in a loop, and then the appropriate values simply assigned to each element of the coefficient array. All of the sums on the right side are unique, and they can be evaluated in the same loop used for the sums appearing on the left. After the coefficient matrix and the right side vector have been evaluated, all that remains is a routine linear system solution.

Although the process works for a polynomial of any degree less than $N_p - 1$, polynomials of large degree are not recommended for most purposes. While a polynomial of higher degree will usually achieve a smaller figure of merit than one of lesser degree, it often will not show the desired smoothing.

Other Least Squares Curve Fits

Exponential Curve Fit

While the polynomial form is widely used and requires only the solution of a system of linear equations to determine the coefficients, there are times when it is appropriate to fit a function other than a polynomial. In some cases, this works quite well, as will be shown in an example to follow. In other cases this does not work well at all. In a section titled "What *not* to compute," Acton (pp. 252 – 253) points out that the problem of fitting the function

$$
f(x) = Ae^{-\alpha x} + Be^{-\beta x}
$$

is quite simple if α and β are known and only A and B are to be determined, but that fitting this same function leads to disaster when $A, B, \alpha,$ and β are to be determined.

Circular Arc Curve Fit

This is an identification problem, where the objective is to provide numerical values for the parameters in a theoretical model of a process. This problem is taken directly from an industrial situation.

In the inspection of nominally circular parts mounted on a shaft, there are basically two questions of interest:

- What is the best estimate for the radius (or diameter) of the part?

- What is the best estimate of the center location?

Inspection measurements are made by mounting the shaft between centers and employing a dial indicator driven by a radial rolling follower, riding against the part circumference. The data consist of two column vectors, $\{\theta_i\}$ and $\{y_i\}$, $i = 1, 2, \ldots N$, representing angular positions of the part and associated indicator readings. The objective is to determine the best answers to the above questions, in a least squares sense.

For a eccentric, perfect circle, the geometry of the measurement system is such that the follower displacement can be described as

$$
\begin{aligned}
Y(\theta) &= -U_c \sin\theta + V_c \cos\theta - R_B - R_F + \sqrt{D} \\
D &= (R_B + R_F)^2 - (U_c \cos\theta + V_c \sin\theta)^2
\end{aligned}
$$

where

Y is the follower displacement measured by the dial indicator

θ is the rotation of the part

R_B is the radius of the part being measured

R_F is the radius of the roller follower

U_c is the center point coordinate as measured perpendicular to the reference line

V_c is the center point coordinate measured parallel to the reference line

The rotation θ is measured from the radial line containing the follower to a reference line fixed on the surface of the part being measured. The eccentricity component U_c is positive in the sense of increasing θ. The second eccentricity component, V_c, is positive radially outward parallel to the reference line.

The measurement system parameter R_F is known, and the parameters to be determined in the curve fitting process are $R_B, U_c,$ and V_c. Note that the function to be fitted to the measured y_i data is $Y(\theta)$, certainly quite far removed from being a simple polynomial expression. It is selected on the basis of an analysis of the ideal system, an eccentric circle.

The errors arising both from the nonideal nature of the actual part and from the measuring system are expressible as

$$\varepsilon_i = Y(\theta_i) - y_i$$

As usual for the least squares process, the object is to minimize a figure of merit consisting of the sum of the squares of the errors,

$$FOM = \sum_{i=1}^{N} \varepsilon_i^2 = \sum_{i=1}^{N} [Y(\theta_i) - y_i]^2$$

The figure of merit is to be minimized with respect to the parameters to be estimated, $R_B, U_c,$ and V_c, so the derivatives with respect to those parameters are developed next. Differentiating with respect to a typical parameter, denoted C, gives

$$\frac{\partial FOM}{\partial C} = 2 \sum_{i=1}^{N} [Y(\theta_i) - y_i] \frac{\partial Y}{\partial C} = 0$$

The partial derivatives of Y with respect to the several specific parameters are as follows:

$$\frac{\partial Y}{\partial R_B} = -1 + \frac{R_B + R_F}{\sqrt{D}}$$

$$\frac{\partial Y}{\partial U_c} = -\sin\theta - \frac{U_c \cos^2\theta}{\sqrt{D}} - \frac{V_c \sin\theta \cos\theta}{\sqrt{D}}$$

$$\frac{\partial Y}{\partial V_c} = \cos\theta - \frac{U_c \sin\theta \cos\theta}{\sqrt{D}} - \frac{V_c \sin^2\theta}{\sqrt{D}}$$

Combining these results gives the system of equations to be solved to obtain the estimates for R_B, U_c and V_c:

$$0 = \sum_{i=1}^{N} \left\{ [Y(\theta_i) - y_i] \left[-1 + \frac{R_B + R_F}{\sqrt{D_i}} \right] \right\}$$

$$0 = \sum_{i=1}^{N} \left\{ [Y(\theta_i) - y_i] \left[-\sin\theta_i - \frac{U_c \cos^2\theta_i}{\sqrt{D_i}} - \frac{V_c \sin\theta_i \cos\theta_i}{\sqrt{D_i}} \right] \right\}$$

$$0 = \sum_{i=1}^{N} \left\{ [Y(\theta_i) - y_i] \left[\cos\theta_i - \frac{U_c \sin\theta_i \cos\theta_i}{\sqrt{D_i}} - \frac{V_c \sin^2\theta_i}{\sqrt{D_i}} \right] \right\}$$

$$D_i = (R_B + R_F)^2 - (U_c \cos\theta_i + V_c \sin\theta_i)^2$$

This is certainly not the usual linear system arising in polynomial least squares curve fits, but this system of equations is amenable to solution by means of the Newton-Raphson method as is discussed in Chapter 9.

References

Acton, F.S., *Numerical Methods That Work*, Harper & Row, 1970.

Chapter 5

Smoothing

Data obtained from physical measurements invariably contains errors. These errors arise from many sources, including human errors, thermally induced electronic errors, truncation errors arising in analog to digital conversion, interaction of the measuring system with the system being observed, errors in the measurement mechanism due to deflections of parts or from friction, etc. Errors that consistently modify the data in the same sense are called *systematic errors*. Systematic errors would include the time-keeping error in a clock with the pendulum length not properly adjusted or the failure to account for the tare weight on a scale. The second major error type is the *random error*. As the name suggests, errors of this sort are random in both magnitude and sign, including, for example, scatter in a target shot by a marksman or the scatter in a dimensional measurement performed by each member of a group of amateurs.

Random errors induce a jaggedness, seen when the data is plotted; whereas, systematic errors are more likely to simply displace the curve. If the measured data is replaced with adjusted data, all falling along a well chosen smooth curve, the random error is reduced. This does nothing, however, about the systematic error. The process of gently nudging the data into a smooth curve is called *smoothing*. LaFora shows that the process of smoothing is equivalent to putting the data through a low-pass filter, which is not surprising. The random error is only evident when it is of relatively high frequency, and the removal of this high frequency noise requires a low-pass filter.

The smoothing processes to be described below are based on the idea of least squares curve fitting, so one might reasonably ask how this differs from the material of Chapter 4. The process described in Chapter 4 is rather aggressive and may smooth excessively, where the processes to be presented here adjust the data in very small steps. As such, it may be necessary to apply the smoothing algorithm repeatedly to achieve the desired result, while the typical least squares curve fit is done only once for a given set of data.

Polynomial Least Squares Smoothing

Suppose that the data to be smoothed consists of N_p data points (x_i, y_i), *where the x-values are evenly spaced,* and no x-values are repeated (except possibly at the ends of a table representing a periodic function). As mentioned before, to apply the polynomial least squares algorithm to the entire data set at once often has the effect of smoothing excessively. The approach employed for the smoothing algorithm is to develop a "local least squares fit," that is, to make a polynomial least squares fit, employing a polynomial of low degree, to the data in the neighborhood of the point to be adjusted. A new fit, centered on the point to be adjusted, is developed for every point except for points close to the end of the table where centering is no longer possible. The number of points used in each fit is denoted as N_f, an *odd number* greater than $1 + N_p$, where N_p is the degree of the polynomial used to make the fit. The value N_f is odd so that the fit may be centered on one particular data point.

Analytical Formulation for the Coefficients

It is convenient to define a value m, where

$$m = \text{int}\left(\frac{N_f}{2}\right) = \text{ greatest integer in } \left(\frac{N_f}{2}\right)$$

A group of data points, centered on x_c and containing N_f points is then described by (x_i, y_i), $c - m \leq i \leq c + m$. For the work to follow, it is more convenient to define a new variable, u_i, such that $u_i = x_i - x_c$ and to consider the data values for the local least squares fit as the set (u_j, v_j), $j = -m, -m+1, \ldots 0 \ldots m$, where

$$\begin{aligned} u_j &= x_i - x_c & j &= i - c \\ v_j &= y_i & -m &\leq j \leq m \end{aligned}$$

From Chapter 4, the equations governing the least squares curve fit are

$$\begin{bmatrix} N_f & \sum_j u_j & \sum_j u_j^2 & \sum_j u_j^3 & \cdots \\ \sum_j u_j & \sum_j u_j^2 & \sum_j u_j^3 & \cdots \\ \sum_j u_j^2 & \sum_j u_j^3 & \cdots \\ \vdots & \vdots \end{bmatrix} \begin{Bmatrix} c_0 \\ c_1 \\ c_2 \\ \vdots \end{Bmatrix} = \begin{Bmatrix} \sum_j v_j \\ \sum_j u_j v_j \\ \sum_j u_j^2 v_j \\ \vdots \end{Bmatrix}$$

which determines a polynomial of the form

$$V(u) = c_0 + c_1 u + c_2 u^2 + \ldots$$

For the particular case at hand, that of evenly spaced data, $u_j = jh$, where h is the interval on u-values (originally on x-values). Thus the equations can be written as

$$\begin{bmatrix} N_f & h\sum_j j & h^2\sum_j j^2 & h^3\sum_j j^3 & \cdots \\ h\sum_j j & h^2\sum_j j^2 & h^3\sum_j j^3 & \cdots \\ h^2\sum_j j^2 & h^3\sum_j j^3 & \cdots \\ \vdots & \vdots \end{bmatrix} \begin{Bmatrix} c_0 \\ c_1 \\ c_2 \\ \vdots \end{Bmatrix} = \begin{Bmatrix} \sum_j v_j \\ h\sum_j j v_j \\ h^2\sum_j j^2 v_j \\ \vdots \end{Bmatrix}$$

or as

$$
\begin{bmatrix}
N_f & \sum\limits_j j & \sum\limits_j j^2 & \sum\limits_j j^3 & \cdots \\
\sum\limits_j j & \sum\limits_j j^2 & \sum\limits_j j^3 & \cdots & \\
\sum\limits_j j^2 & \sum\limits_j j^3 & \cdots & & \\
\vdots & \vdots & & &
\end{bmatrix}
\underbrace{\begin{Bmatrix} c_0 \\ hc_1 \\ h^2 c_2 \\ \vdots \end{Bmatrix}}_{(N_p+1\times 1)}
$$
$$\underbrace{\phantom{\begin{bmatrix}N_f\end{bmatrix}}}_{(N_p+1 \times N_p+1)}$$

$$
= \underbrace{\begin{Bmatrix} \sum\limits_j v_j \\ \sum\limits_j j v_j \\ \sum\limits_j j^2 v_j \\ \vdots \end{Bmatrix}}_{(N_p+1\times 1)}
$$

$$
= \underbrace{\begin{bmatrix}
1 & 1 & 1 & \cdots \\
-m & 1-m & 2-m & \cdots \\
m^2 & (1-m)^2 & (2-m)^2 & \cdots \\
\vdots & \vdots & &
\end{bmatrix}}_{((N_p+1)\times(2m+1))}
\underbrace{\begin{Bmatrix} v_{-m} \\ v_{1-m} \\ \vdots \\ v_0 \\ \vdots \\ v_m \end{Bmatrix}}_{(2m+1\times 1)}
$$

This is simply a system of linear equations, the usual polynomial least squares curve fitting equations in modified form, and can be readily solved by multiplying by the inverse of the coefficient matrix. The result will be values for the coefficients, $col\left(c_0, hc_1, h^2c_2, \ldots\right)$. Note that the sums of odd powers, such as $\sum j$ or $\sum j^3$, will all be zero. Thus only about half of the elements in the coefficient matrix will be nonzero.

The objective for all of this has been to compute a smoothed value for the center point, and perhaps for the points before and after the center, if the center point is near the beginning or the end of the data set, respectively. These will be expressed as

$$
\begin{aligned}
V_j &= c_0 + c_1 u_j + c_2 u_j^2 + \ldots \\
&= c_0 + jhc_1 + j^2 h^2 c_2 + \ldots \\
&= c_0 + j\left(hc_1\right) + j^2\left(h^2 c_2\right) + \ldots
\end{aligned}
$$

The full set of V_j determined in any single fit is given by

$$
\{V_j\} = \begin{bmatrix} 1 & -m & m^2 & \cdots \\ 1 & 1-m & (1-m)^2 & \\ 1 & 2-m & (2-m)^2 & \\ \vdots & \vdots & \vdots & \end{bmatrix} \begin{Bmatrix} c_0 \\ hc_1 \\ h^2 c_2 \\ \vdots \end{Bmatrix}
$$

$$
= [M(m)] \begin{Bmatrix} c_0 \\ hc_1 \\ h^2 c_2 \\ \vdots \end{Bmatrix}
$$

where $[M(m)]$ is the coefficient matrix indicated. This same matrix appeared earlier in transposed form in the equations for the least squares fit coefficients. The vector on the right is exactly the result obtained earlier when the equations for the least squares fit coefficients were solved. That earlier result may be substituted to obtain

$$
\{V_j\} = [M(m)] \begin{bmatrix} N_f & \sum_j j & \cdots \\ \sum_j j & \sum_j j^2 & \\ \vdots & \vdots & \end{bmatrix}^{-1} [M(m)]^t \begin{Bmatrix} v_{-m} \\ v_{1-m} \\ \vdots \\ v_0 \\ \vdots \\ v_m \end{Bmatrix}
$$

$$
= [f(m)] \{v_i\}
$$

Note that this expresses each of the smoothed values, V_j as a linear combination of the original data values, v_i. The matrix $[f(m)]$ is the indicated product of three matrices, and the elements of $[f(m)]$ are multipliers for the v_i, each row serving to compute one adjusted value, V_j. Note also that the third matrix in the product is the transpose of the first. Thus this matrix only has to be evaluated once. With this formulation, it is easy to determine the required coefficients for any number of data points fitted with a polynomial of any degree less than $N_f - 1$. While there are not a great many cases of interest, the consistency of this formulation with computer implementation tends to eliminate errors that might be induced in a manual calculation.

Cases of Interest

Linear Fit

For a linear fit using three data points, the smoothing equations are

$$
\begin{aligned}
V_{-1} &= (5v_{-1} + 2v_0 - v_1)/6 \\
V_0 &= (2v_{-1} + 2v_0 + 2v_1)/6 \\
V_1 &= (-v_{-1} + 2v_0 + 5v_1)/6
\end{aligned}
$$

For a linear fit using five data points, the smoothing equations are

$$
\begin{aligned}
V_{-2} &= (6v_{-2} + 4v_{-1} + 2v_0 + 0v_1 - 2v_2)/10 \\
V_{-1} &= (4v_{-2} + 3v_{-1} + 2v_0 + 1v_1 + 0v_2)/10 \\
V_0 &= (2v_{-2} + 2v_{-1} + 2v_0 + 2v_1 + 2v_2)/10 \\
V_1 &= (0v_{-2} + v_{-1} + 2v_0 + 3v_1 + 4v_2)/10 \\
V_2 &= (-2v_{-2} + 0v_{-1} + 2v_0 + 4v_1 + 6v_2)/10
\end{aligned}
$$

Note the top-to-bottom symmetry in this list and in the one above it. The coefficients for computing V_2 are the same group of numbers as those used to compute V_{-2}, taken in reverse order. To shorten the presentations to follow, only the first half of each table will be presented.

For a linear fit using seven data points, the smoothing equations are

$$
\begin{aligned}
V_{-3} &= (13v_{-3} + 10v_{-2} + 7v_{-1} + 4v_0 + v_1 - 2v_2 - 5v_3)/28 \\
V_{-2} &= (10v_{-3} + 8v_{-2} + 6v_{-1} + 4v_0 + 2v_1 + 0v_2 - 2v_3)/28 \\
V_{-1} &= (7v_{-3} + 6v_{-2} + 5v_{-1} + 4v_0 + 3v_1 + 2v_2 + 1v_3)/28 \\
V_0 &= (4v_{-3} + 4v_{-2} + 4v_{-1} + 4v_0 + 4v_1 + 4v_2 + 4v_3)/28 \\
&\vdots
\end{aligned}
$$

Quadratic Fit

For a quadratic fit using five data points, the smoothing equations are

$$
\begin{aligned}
V_{-2} &= (31v_{-2} + 9v_{-1} - 3v_0 - 5v_1 + 3v_2)/35 \\
V_{-1} &= (9v_{-2} + 13v_{-1} + 12v_0 + 6v_1 - 5v_2)/35 \\
V_0 &= (-3v_{-2} + 12v_{-1} + 17v_0 + 12v_1 - 3v_2)/35 \\
&\vdots
\end{aligned}
$$

For a quadratic fit using seven data points, the smoothing equations are

$$
\begin{aligned}
V_{-3} &= (32v_{-3} + 15v_{-2} + 3v_{-1} - 4v_0 - 6v_1 - 3v_2 + 5v_3)/42 \\
V_{-2} &= (15v_{-3} + 12v_{-2} + 9v_{-1} + 6v_0 + 3v_1 + 0v_2 - 3v_3)/42 \\
V_{-1} &= (3v_{-3} + 9v_{-2} + 12v_{-1} + 12v_0 + 9v_1 + 3v_2 - 6v_3)/42 \\
V_0 &= (-4v_{-3} + 6v_{-2} + 12v_{-1} + 14v_0 + 12v_1 + 6v_2 - 4v_3)/42 \\
&\vdots
\end{aligned}
$$

Cubic Fit

For a cubic fit using five data points, the smoothing equations are

$$
\begin{aligned}
V_{-2} &= \left(207v_{-2} + 12v_{-1} - 18v_0 + 12v_1 - 3v_2\right)/210 \\
V_{-1} &= \left(12v_{-2} + 162v_{-1} + 72v_0 - 48v_1 + 12v_2\right)/210 \\
V_0 &= \left(-18v_{-2} + 72v_{-1} + 102v_0 + 72v_1 - 18v_2\right)/210 \\
&\vdots
\end{aligned}
$$

For a cubic fit using seven data points, the smoothing equations are

$$
\begin{aligned}
V_{-3} &= \left(39v_{-3} + 8v_{-2} - 4v_{-1} - 4v_0 + 1v_1 + 4v_2 - 2v_3\right)/42 \\
V_{-2} &= \left(8v_{-3} + 19v_{-2} + 16v_{-1} + 6v_0 - 4v_1 - 7v_2 + 4v_3\right)/42 \\
V_{-1} &= \left(-4v_{-3} + 16v_{-2} + 19v_{-1} + 12v_0 + 2v_1 - 4v_2 + 1v_3\right)/42 \\
V_0 &= \left(-4v_{-3} + 6v_{-2} + 12v_{-1} + 14v_0 + 12v_1 + 6v_2 - 4v_3\right)/42 \\
&\vdots
\end{aligned}
$$

It should be mentioned in passing that these results are in agreement with Hildebrand but not quite in accord with those given by LaFora. (It seems likely that the coefficients given by LaFora were obtained by hand calculation and are therefore subject to random arithmetic errors.) The results given here were all obtained using the general formulation presented earlier.

Application

At this point, the major remaining question is, "Just exactly how do we implement the smoothing process?" Recall that the original data set consisted of N_p pairs, where we assume that N_p is a fairly large number. We may very well want to plot the original data set simply to get some idea what the most appropriate curve fit form may be where we will be choosing between linear, quadratic, or cubic forms. It is also necessary to select the number of points to be used for each fit, an odd number. There is not a single correct way to make these choices. Smoothing is definitely an art more than a science. Assuming that these decisions have been made, the process for one smoothing pass is

$$
\left.
\begin{aligned}
Y_0 &= v_{-m} \\
Y_1 &= v_{1-m} \\
Y_2 &= v_{2-m} \\
&\vdots \\
Y_m &= v_0
\end{aligned}
\right\} \quad \text{From the first group fitted}
$$

$$Y_{m+1} = v_0$$
$$\vdots$$
$$Y_{N_{dp}-m-1} = v_0$$

} From the many groups away from both ends

$$Y_{N_{dp}-m} = v_0$$
$$\vdots$$
$$Y_{N_{dp}-1} = v_{m-1}$$
$$Y_{N_{dp}} = v_m$$

} From the last group fitted

Thus the first group fitted provides the first $m + 1$ adjusted values. This is then followed by most of the adjusted values computed simply at the center of the group until the center point approaches the upper end of the table. The last $m+1$ adjusted values are all determined from the final fit.

Strictly speaking, all of the adjusted values should be determined based on the previous set of values. In practice, however, it is common to compute the values "in place" which leads to a new center point value expressed as

$$V_0 = (-4V_{-3} + 6V_{-2} + 12V_{-1} + 14v_0 + 12v_1 + 6v_2 - 4v_3)/42$$

based, for example, on a quadratic fit to seven data points. Note that the v-values with negative subscripts have all been previously adjusted. This approach appears justified in that

- The adjusted values represent an artistic revision of the original data, not the data itself

- The computation is faster and more compact in this form

It is also a common practice to use repeated smoothing passes, depending on the needs of the situation. In some cases, five smoothing passes may be adequate, but there are also cases where 50 passes is more appropriate, and in a few extreme cases, this writer has used as many as 500 smoothing passes to clean up data for which theory predicts a very smooth curve.

Example

In order to make use of the smoothing algorithms given above, computer code for that purpose is required. The block of code listed below incorporates the least squares cubic polynomial fit to five data points at a time. Note that as mentioned above, the recently adjusted values are used in the adjustment of the values immediately following. The data is passed to the subroutine as v(), and it is returned in the same array. Because five points are to be used for each fit, the first two points in the adjusted data must be determined using the formulae for V_{-2} and V_{-1}. This is then followed by the use of the formula for V_0 for data index values from 2 to $N_{dp} - 2$. Finally, the last two data points are adjusted using the equations for V_1 and V_2.

43

Example Smoothing Subroutine

```
sub smooth3_5(v(),Ndp)
    ! This routine uses cubic least squares
    ! smoothing applied to five points
    ! at a time
    i=2
    v(i-2)=(207*v(i-2)+12*v(i-1)-18*v(i)&
&    +12*v(i+1)-3*v(i+2))/210
    v(i-1)=(12*v(i-2)+162*v(i-1)+72*v(i)&
&    -48*v(i+1)+12*v(i+2))/210
    for i=2 to Ndp-2
        v(i)=(-18*v(i-2)+72*v(i-1)+102*v(i)&
&        +72*v(i+1)-18*v(i+2))/210
    next i
    i=Ndp-2
    v(i+1)=(12*v(i-2)-48*v(i-1)+72*v(i)&
&    +162*v(i+1)+12*v(i+2))/210
    v(i+2)=(-3*v(i-2)+12*v(i-1)-18*v(i)&
&    +12*v(i+1)+207*v(i+2))/210
end sub
```

In order to test the operation of the subroutine smooth3_5, a numerical experiment is useful. A set of perfectly smooth data is generated using the definition

$$Y_{\text{orig}}(i) = \sin\left(\frac{\pi x_i}{2}\right) + x_i \qquad x_i = i/100 \qquad i = 0, 1, 2, \ldots 150$$

Smoothness is assured because of the nature of the function defining Y_{orig}. Then, for the purposes of the experiment, this data set is corrupted by adding a randomly distributed value, r, lying in the range $-0.1 \leq r \leq 0.1$, The corrupted data set is called $Y_{\text{rnd}}(i)$. The subroutine smooth3_5 is then applied to the corrupted data 100 times, i.e., the corrupted data is passed through the subroutine 100 times. The result is called $Y_{\text{sm}}(i)$. A sample of the results expressed as differences is shown in the following table. It appears that Y_{sm} is somewhat closer to Y_{orig} than to Y_{rnd}. Equally important, notice that Y_{sm} does not differ by very much from the corrupted data, Y_{rnd}, indicating that even after 100 passes through the smoothing routine, only very small shifts have been made in the data.

Results Comparison for Smoothing Experiment

Index	Y_rnd-Y_sm	Y_sm-Y_orig	Y_rnd-Y_orig
0	+.01497	−.00527	+.0097
1	−.05901	−.02533	−.08435
2	+.06829	−.03071	+.03758
3	−.00626	−.02669	−.03295
4	+.10861	−.01822	+.09038
5	+.07413	−.00939	+.06475

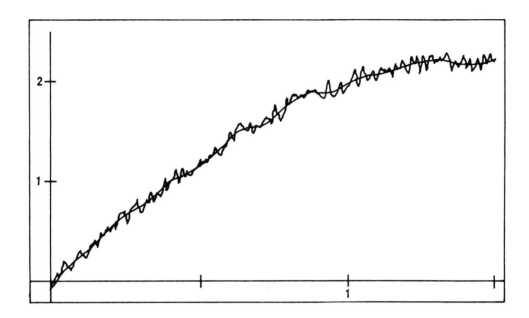

Figure 1: Plots of the corrupted data and the smoothed result.

18	−.01242	+.0092	−.00322
19	+.03941	+.01637	+.05578
20	−.03221	+.02328	−.00894
47	−.05669	−.03774	−.09443
48	−.00605	−.03866	−.04471
49	−.01521	−.03713	−.05234
90	−.01274	+.00621	−.00652
91	−.03671	−.0074	−.04411
92	−.04638	−.0201	−.06648
128	+.03824	+.03214	+.07038
129	−.02612	+.03362	+.0075
130	−.01826	+.03303	+.01477
147	−.00553	−.00601	−.01155
148	+.06369	+.00248	+.06617
149	−.02062	+.01324	−.00738
150	+.00398	+.02696	+.03093

The degree of smoothing achieved is much more evident in Figure 1. This figure shows the corrupted data and the smoothed data, plotted together. The original, uncorrupted data is not shown. It is evident that there is still a little bumpiness in the data that was induced by the random error term added, but the smoothed data is clearly just that: it is smoothed!

The significance of smoothing before undertaking a process such as numerical differentiation of a data set is apparent in this figure. While the slope of the smoothed curve is rarely exactly the same as that of the original (smooth) data, it deviates far less than the slope of the corrupted data, almost everywhere.

References

Hildebrand, F.B., *Introduction to Numerical Analysis*, McGraw-Hill, 1956.

LaFora, R., *Computer Methods for Science and Engineering*, Hayden, 1973.

Chapter 6

Interpolation and Differentiation of Tabular Functions

In dealing with a tabular function known only as a list of values, we often need to interpolate to obtain an estimate of the function value for an argument not in the table. Such an estimate is invariably based, at least implicitly, on the idea that the function is smooth and that a smooth curve joining the tabulated values is a reasonable estimate for the function at the intermediate locations. It must be noted that *this is strictly an assumption.*

A second need that may also arise is the need to estimate values for the derivative of the function, either at the tabulation points or at intermediate points. Again, this is based on the assumption of smoothness mentioned above. Traditionally, differentiation of tabular functions has been considered taboo; the results were often badly in error. The availability of high speed digital computation now makes practical some methods that were previously too time consuming to be useful. Even so, such differentiation needs to be done with care and the results critically examined.

Lagrange Interpolating Polynomial

Provided that the function is truly smooth, both of these needs can be addressed by means of the Lagrange interpolating polynomial. The process is essentially this:

- A polynomial is exactly fitted to several points around the location where interpolation and/or differentiation is required.

- This polynomial is then differentiated if required, the polynomial and its derivative are evaluated numerically, and the results are offered as approximations to the (unknown) function value and function derivative.

Consider the tabular function defined below:

$$
\begin{array}{ccccccccc}
x: & x_1 & x_2 & \cdots & x_p & x_{p+1} & \cdots & x_{q-1} & x_q & \cdots \\
y: & y_1 & y_2 & \cdots & y_p & y_{p+1} & \cdots & y_{q-1} & y_q & \cdots
\end{array}
$$

where the x_i need not be evenly spaced. In fact, *there is no advantage to having even or uneven spacing* when using the Lagrange polynomial. Now suppose that the question of interest is to obtain estimates for $y(x)$ and $y'(x) = \frac{dy}{dx}\big|_x$, where x may or may not be one of the tabulated arguments.

One of the first problems is to locate the appropriate part of the table by some sort of search. Let us here assume that this has been accomplished and that the $n+1$ points with subscripts p through q are those selected as the basis for the interpolation. It is desirable that x be roughly centered in the interval $[x_p, x_q]$, although the algorithm to be developed below will work to some degree even if the point x is not well centered in the data set.

Because $n+1$ data points are to be used, a polynomial of degree n can be exactly fitted to these points. Such a polynomial is often written as

$$P(x) = c_0 + c_1 x + c_2 x^2 + \cdots + c_n x^n$$

but this polynomial can also be written in factored form as

$$P(x) = K(x - x_1)(x - x_2)(\cdots)(x - x_n)$$

where x_1, x_2, $\cdots$ x_n are the zeros of $P(x) = 0$.

Before addressing the problem of finding a suitable polynomial to meet all the requirements for the interpolation, consider the construction of a polynomial that will be zero at all but one of the tabulation points in the interval $[x_p, x_q]$. Let this polynomial be denoted by $P_k(x)$, and it will be zero at x_p, x_{p+1}, $\cdots$ x_q, except that it will be nonzero at x_k. Because the zeros of the polynomial are specified, constructing such a polynomial is an easy matter:

$$P_k(x) = (x - x_p)(x - x_{p+1})(\cdots)(x - x_{k-1})(x - x_{k+1})(\cdots)(x - x_q)$$

Note that there is no factor $(x - x_k)$ involved in $P_k(x)$. This is a polynomial of degree n because it involves exactly n factors, each containing an x. This form could be multiplied by any constant, but that will not alter the fact that it has a zero at each tabulation point in the interval $[x_p, x_q]$, except at x_k.

If the polynomial just constructed above is evaluated at $x = x_k$, the result is:

$$P_k(x_k) = (x_k - x_p)(x_k - x_{p+1})(\cdots)(x_k - x_{k-1})(x_k - x_{k+1})(\cdots)(x_k - x_q)$$

Evidently then, a polynomial of degree n passing through (x_k, y_k) and having a zero value at each of the other tabulation points in the interval $[x_p, x_q]$ can be constructed as

$$
\begin{aligned}
\tilde{P}_k(x) &= y_k \frac{P_k(x)}{P_k(x_k)} \\
&= y_k \frac{(x - x_p)(x - x_{p+1})(\cdots)(x - x_{k-1})(x - x_{k+1})(\cdots)(x - x_q)}{(x_k - x_p)(x_k - x_{p+1})(\cdots)(x_k - x_{k-1})(x_k - x_{k+1})(\cdots)(x_k - x_q)}
\end{aligned}
$$

The next step is to extend this to a polynomial passing through all of the tabulation points. Recall that if we add several polynomials of degree n, the result will still be a polynomial of degree n. Thus, consider the sum

$$
\begin{aligned}
P_L(x) &= \sum_{k=p}^{k=q} \tilde{P}_k(x) \\
&= \sum_{k=p}^{k=q} y_k \frac{(x - x_p)(x - x_{p+1})(\cdots)(x - x_{k-1})(x - x_{k+1})(\cdots)(x - x_q)}{(x_k - x_p)(x_k - x_{p+1})(\cdots)(x_k - x_{k-1})(x_k - x_{k+1})(\cdots)(x_k - x_q)}
\end{aligned}
$$

This is the Lagrange polynomial $P_L(x)$. While somewhat formidable in appearance, it is simply the summation of the simple polynomials developed earlier, each of which passed through one tabulation point and was zero at the other tabulation points in the interval $[x_p, x_q]$. Despite its appearance, it is not difficult to program this expression for computer evaluation at any value x. An example of the necessary code will follow below.

One of the questions that is often raised at this point is, "OK, but is that the correct polynomial? Is there possibly another polynomial of the same degree and passing through those same points that should be used instead?" The answers to these questions are simply, "Yes and No, respectively." Although we may put the polynomial into different forms, the polynomial itself is unique: there is one and only one polynomial with the stated properties.

In the code fragment below, the data, x() and y(), is assumed to be already in place. The evaluation of the Lagrange polynomial, $P_L(x)$, is performed in the for-next loop on the index k, essentially parallel to the development given previously. The evaluation location is denoted as xx in the program. Each polynomial is evaluated separately in the for-next loop on s where the numerator and denominator polynomials, pin and pid, are evaluated. The contribution of each polynomial is included in the summation in the line that reads f = f + y(k)*Lk. The remainder of the example deals with the evaluation of the derivative estimate to be considered following the example code for Lagrange polynomial interpolation and differentiation.

Subroutine for Lagrange Polynomial
Interpolation & Differentiation

```
sub LagrPoly(xx,x(),y(),Pdeg,Nv,f,fp)
   ! input information
   ! xx     Evaluation point
   ! x()    Data point locations
   ! y()    Data values
   ! Pdeg   Degree of interpolating polynomial
   ! Nv     Largest index in each of x() and y()
   ! f      Returned function estimate
   ! fp     Returned derivative estimate
   ! Starting index assumed to be zero
```

49

```
    ! if pdeg is odd ...
    if pdeg<>2*(int(pdeg/2)) then&
&       shift=int((pdeg-1)/2)
    ! if pdeg is even ...
    if pdeg=2*(int(pdeg/2)) then&
&       shift=int(pdeg/2)

    ! Search for interval
    for isr=0 to Nv-1
       if (x(isr)-xx)*(x(isr+1)-xx)<=0 then
         ! near beginning of table
         p=max(isr-shift,0)
         exit for
       end if
    next isr
    ! near end of the table
    q=min(p+Pdeg,Nv)
    p=q-Pdeg
    f=0
    fp=0
    for k=p to q
       xk=x(k)
       pin=1  ! Initialize numerator product
       pid=1  ! Initialize denominator product
       for i=p to q
         if i<>k then
           pid=pid*(xk-x(i))
           pin=pin*(xx-x(i))
         end if
       next i
       ! Kth Lagrange polynomial
       Lk=pin/pid
       ! Accumulating function estimate
       f=f+y(k)*Lk

       sumk=0
       for j=p to q
         if j<>k then
           pr=1
           for r=p to q
             if r<>k and r<>j then&
&                pr=pr*(xx-x(r))
           next r
           sumk=sumk+pr
```

```
      end if
   next j
   ! d(Lk)/dx = sumk/pid
   dLkdx=sumk/pid
   ! Accumulating derivative estimate
   fp=fp+y(k)*dLkdx

 next k
end sub
```

Considering the form for the individual interpolating polynomials, $\tilde{P}_k(x)$, note that

- the y_k are simply constants, and

- the denominator is simply a constant.

The only place the variable x appears is in the several factors appearing in the numerator. Evidently then, $\tilde{P}_k(x)$ must be differentiated as a product. Looking only at that part of the whole expression for $\tilde{P}_k(x)$, the derivative is

$$\frac{d[(x-x_p)(x-x_{p+1})(\cdots)(x-x_{k-1})(x-x_{k+1})(\cdots)(x-x_q)]}{dx}$$

$$= (x-x_{p+1})(\cdots)(x-x_{k-1})(x-x_{k+1})(\cdots)(x-x_q)$$
$$+ (x-x_p)(\cdots)(x-x_{k-1})(x-x_{k+1})(\cdots)(x-x_q)$$
$$+ \ldots$$
$$+ (x-x_p)(x-x_{p+1})(\cdots)(x-x_{k+1})(\cdots)(x-x_q)$$
$$+ (x-x_p)(x-x_{p+1})(\cdots)(x-x_{k-1})(\cdots)(x-x_q)$$
$$+ \cdots$$
$$+ (x-x_p)(x-x_{p+1})(\cdots)(x-x_{k-1})(x-x_{k+1})(\cdots)$$

$$= \sum_{j=p}^{j=q} \prod_{\substack{r=p \\ r\neq k \\ r\neq j}}^{r=q} (x-x_r)$$

so that the derivative of the entire Lagrange polynomial becomes

$$\frac{dP_L(x)}{dx} = \sum_{k=p}^{k=q} \frac{y_k}{\displaystyle\prod_{\substack{s=p \\ s\neq k}}^{s=q}(x_k-x_s)} \times \sum_{\substack{j=p \\ j\neq k}}^{j=q} \prod_{\substack{r=p \\ r\neq k \\ r\neq j}}^{r=q} (x-x_r)$$

In the example code, the quantity sumk is the value of everything above following the $\times$ for each value of k.

The process set out above could in principle be extended directly to higher derivatives by differentiating the products again. The bookkeeping in such a process becomes laborious, however, and there is an alternate approach available.

51

Suppose that the process described above is implemented in a subroutine called
LagrPoly(xx,x(),y(),yy,dydx) where

    xx   = evaluation location
    x()   = array of tabulation locations
    y()   = arrary of tabulated values
    yy   = interpolation result
    dydx   = polynomial derivative at xx

The sequence of calls

```
for i=1 to N
    call LagrPoly(xx,x(),y(),yy,dydx)
    yp(i)=dydx
    next i

    for i=1 to N
        call LagrPoly(xx,x(),yp(),yy,dydx)
        ypp(i)=dydx
    next i

    for i=1 to N
        call LagrPoly(xx,x(),ypp(),yy,dydx)
        yppp(i)=dydx
    next i
```

will develop the arrays

    yp   = tabulated first derivative estimates
    ypp   = tabulated second derivative estimates
    yppp   = tabulated third derivative estimates

The accumulated rounding error increases with each application of the Lagrange
polynomial, so there are limits to this process. Even so, this has been found to be
a highly useful technique in such areas as cam design and inspection.

Newton's Divided Differences

The other prominent method for interpolation of tabular functions is Newton's
Divided Difference interpolating polynomial. As with the Lagrange polynomial, it is
not required that the data be evenly spaced. The theory for this method is developed
below. Without even spacing of the data, Newton's divided difference form does not
lend itself to approximating derivatives, although it can certainly be done if one is
willing to expend the effort. This will not be undertaken here.

It is assumed that the process discussed previously regarding the selection of an
appropriate group of data points on which to base the interpolation has been carried
out, and that there is in hand a set of $n+1$ data points, (x_p, y_p), (x_{p+1}, y_{p+1}) $\cdots$
(x_q, y_q) to be interpolated using a Newton divided difference polynomial of degree
n.

Let $P_{L(p,m)}(x)$ denote a Lagrange polynomial fitted at the points $p, p+1, \ldots m$, and correspondingly, $P_{L(p+1,m+1)}(x)$ is a Lagrange polynomial fitted at points $p+1$, $p+2, \ldots m+1$. Each of these will be of degree $m-p$, but each fits a different set of points. The Lagrange polynomial fitted at points $p, p+1, \ldots m, m+1$ will be

$$P_{L(p,m+1)}(x) = \frac{(x_{m+1} - x) P_{L(p,m)}(x) + (x - x_p) P_{L(p+1,m+1)}(x)}{x_{m+1} - x_p}$$

Note that the first term in the numerator has a zero at $x_p, x_{p+1}, \ldots x_m$ because of the factor $P_{L(p,m)}$, and it has a zero at x_{m+1} due to the first factor. In a similar fashion, the second term also has zeroes at each of $x_p, x_{p+1}, \ldots x_m$. At $x = x_j$ where $p+1 \le j \le m$, this gives

$$P_{L(p,m+1)}(x_j) = \frac{(x_{m+1} - x_j) y_j + (x_j - x_p) y_j}{x_{m+1} - x_p} = \frac{(x_{m+1} - x_p) y_j}{x_{m+1} - x_p} = y_j$$

and at $x = x_p$ this is

$$P_{L(p,m+1)}(x_p) = \frac{(x_{m+1} - x_p) y_p + (x_p - x_p) P_{L(p+1,m+1)}(x_p)}{x_{m+1} - x_p} = y_p$$

and finally, at $x = x_{m+1}$

$$P_{L(p,m+1)}(x_{m+1}) = \frac{(x_{m+1} - x_{m+1}) P_{L(p,m)}(x_{m+1}) + (x_{m+1} - x_p) y_{m+1}}{x_{m+1} - x_p} = y_{m+1}$$

Now suppose that the two terms are each expanded as

$$\frac{(x_{m+1} - x) P_{L(p,m)}(x)}{x_{m+1} - x_p} = \frac{(x_{m+1} - x)(\alpha_{m-p} x^{m-p} + \alpha_{m-p-1} x^{m-p-1} + \ldots)}{x_{m+1} - x_p}$$

$$\frac{(x - x_p) P_{L(p+1,m+1)}(x)}{x_{m+1} - x_p} = \frac{(x - x_p)(\beta_{m-p} x^{m-p} + \beta_{m-p-1} x^{m-p-1} + \ldots)}{x_{m+1} - x_p}$$

while the expanded polynomial $P_{L(p,m+1)}(x)$ is of the form

$$P_{L(p,m+1)}(x) = \gamma_{m+1-p} x^{m+1-p} + \gamma_{m-p} x^{m-p} + \cdots$$

Most of the relations between the coefficients γ_i, α_j, and β_k are rather complicated, but the first one, the relation between the coefficients of x^{m+1-p}, is relatively simple:

$$\gamma_{m+1-p} = \frac{\beta_{m-p} - \alpha_{m-p}}{x_{m+1} - x_p}$$

This form is called a divided difference (for rather obvious reasons!), and it tells us that the leading coefficient in the new polynomial is obtained in this particular manner from the leading coefficients of the two previous polynomials.

The Newton interpolating polynomial has the form:

$$
\begin{aligned}
P_n(x) &= c_0 + c_1(x - x_p) + c_2(x - x_p)(x - x_{p+1}) + \ldots \\
&\quad + c_n(x - x_p)(\cdots)(x - x_q) \\
&= c_0 + (x - x_p)\{c_1 + (x - x_{p+1})[c_2 + (x - x_{p+2})(\cdots)]\} \\
&= \ldots \{c_{n-2} + (x - x_{q-1})[c_{n-1} + c_n(x - x_q)]\}
\end{aligned}
$$

This is a sum of polynomials of all degrees up to and including n, a sum consisting of a constant, a linear term, a quadratic term, ... ending with a polynomial of degree n. Note that x_p is a root of all of the polynomials after the constant. Similarly, x_{p+1} is a root of all the polynomials after the linear term, and so forth, until x_q is a root only of the last polynomial.

If an (extremely short) polynomial is to be fitted to the first data point, it is

$$
P_{(p)} = y_p = c_0 \qquad \text{a constant}
$$

Similar fits to the second and third points are given by

$$
\begin{aligned}
P_{(p+1)} &= y_{p+1} = c_0' \qquad \text{a constant} \\
P_{(p+2)} &= y_{p+2} = c_0'' \qquad \text{a constant}
\end{aligned}
$$

where the $'$ and $''$ denote different polynomials, not derivatives. Following the procedure outlined above, polynomials passing through the first and second points (left column) and through the second and third points (right column)

$$
P_{(p,p+1)} = c_0 + c_1 x \qquad P_{(p+1,p+2)} = c_0' + c_1' x
$$

$$
c_1 = \frac{c_0' - c_0}{x_{p+1} - x_p} \qquad\qquad c_1' = \frac{c_0'' - c_0'}{x_{p+2} - x_{p+1}}
$$

with c_1 and c_1' determined according to the divided difference form above. A quadratic polynomial passing through all three points is given by

$$
\begin{aligned}
P_{(p,p+2)} &= c_0 + c_1 x + c_2 x^2 \\
c_2 &= \frac{c_1' - c_1}{x_{p+2} - x_p}
\end{aligned}
$$

If a cubic polynomial passing through four data points were required, then the process would have begun by fitting constants to each of those points, then developing three linear equations, two quadratic equations, and finally, a single cubic equation. The extension to polynomials of yet higher degree is evident and is summarized in the following table:

54

Divided Difference Table

$c_0 = y_i$	$c_1 = D^1$ = First Diff	$c_2 = D^2$ = Second Diff		$\cdots$ $c_n = D^n$
y_p	$D^1_{(0,1)} = \frac{y_{p+1}-y_p}{x_{p+1}-x_p}$	$D^2_{(0,2)} = \frac{D^1_{(1,2)}-D^1_{(0,1)}}{x_{p+2}-x_p}$		$\cdots$ $D^n_{(0,n)}$
y_{p+1}	$D^1_{(1,2)} = \frac{y_{p+2}-y_{p-1}}{x_{p+2}-x_{p+1}}$	$D^2_{(1,3)} = \frac{D^1_{(2,3)}-D^1_{(1,2)}}{x_{p+3}-x_{p+1}}$	$\cdots$	
y_{p+2}	$D^1_{(2,3)} = \frac{y_{p+3}-y_{p+2}}{x_{p+3}-x_{p+2}}$	$D^2_{(2,4)} = \frac{D^1_{(3,4)}-D^1_{(2,3)}}{x_{p+4}-x_{p+2}}$	$\cdots$	
y_{p+3}	$D^1_{(3,4)} = \frac{y_{p+4}-y_{p+3}}{p+4-x_{p+x3}}$	$D^2_{(3,5)} = \frac{D^1_{(4,5)}-D^1_{(3,4)}}{x_{p+5}-x_{p+3}}$	$\cdots$	
$\vdots$		$\vdots$		
y_{q-2} $D^1_{(n-2,n-1)} = \frac{y_{q-1}-y_{q-2}}{x_{q-1}-x_{q-2}}$		$D^2_{(n-2,n)} = \frac{D^1_{(n-1,n)}-D^1_{(n-2,n-1)}}{x_q-x_{q-2}}$		
y_{q-1} $D^1_{(n-1,n)} = \frac{y_q-y_{q-1}}{x_q-x_{q-1}}$				
y_q				

The tabulation begins with the first column listing y-values, largely a hold over from the days of manual computation but also a reminder that this is c_0 for each of the one-term fits that start the whole process. The first divided differences are listed in the second column, and this list is one element shorter than those before it. The third column lists the second divided differences and is one element shorter than the previous column. The final column consists of a single element only. The elements of the top row are the coefficients for the Newton interpolating polynomial. It is not difficult to devise a computer program for the rapid evaluation of all the elements in the difference table, as shown below.

Demonstration of Newton's Interpolating Polynomial

```
!                        NewtIntp.Tru
! Newton Interpolating Polynomial for F(x)
option nolet
option base 0
library "graphlib"
dim x(100),y(100),xx(100),yy(100)
dim ndd(0,0),ff(0),et(0)
clear
print #0: "Do you want"
print #0: " (1)  Evenly spaced input data"
print #0: " (2)  Unevenly spaced input data"
```

```
print #0: "Enter number for your choice"
input an$
if an$="1" then   z=0
if an$="2" then   z=1
Pdeg=7       ! degree of interpolating polynomial
mat redim ndd(pdeg+1,pdeg+1)
mat redim ff(pdeg+1),et(pdeg+1)
call dataset   ! generate the data set
! Data set values are indexed from 0 to 25
! This provides the required value of Nv (=25)

for ix=0 to 100   ! Develop interpolated curves
    xx(ix)=ix/4   ! Interpolation locations
    xxx=xx(ix)
    call nddt
    call newt
    yy(ix)=fval
next ix

! Plot data set
clear
xmin=-2
xmax=25
ymin=-15
ymax=25
set window xmin,xmax,ymin,ymax
box lines xmin,xmax,ymin,ymax
call axes
call ticks(1,2)
! Tick marks on horizontal axis
for i=5 to 20 step 5
    i$=str$(i)
    plot text, at i-.2,-1.6: i$
next i
! Tick marks on vertical axis
for j=-10 to 20 step 10
    j$=str$(j)
    plot text, at -1.5,j-.3: j$
next j
for i=0 to 25          ! Plot data points with +
    plot x(i)-.2,y(i);
    plot x(i)+.2,y(i)
    plot x(i),y(i)-.4;
    plot x(i),y(i)+.4
next i
```

```
txt1$="Evenly Spaced Data Points Plotted"
txt2$="Unevenly Spaced Data Points Plotted"
txt3$="Function Values Plotted with a +"
txt4$="Results from NewtPoly.Tru"
txt5$="Interpolated Function Curve Plotted"
plot text, at 10,-8: txt3$
plot text, at 4,22: txt4$
if z=0 then plot text, at 4,20: txt1$
if z=1 then plot text, at 4,20: txt2$
get key xxxx

for ix=0 to 100 ! Plot the interpolated curve
    plot xx(ix),yy(ix);
next ix
plot
plot text, at 4,18: txt5$
get key zzz

sub nddt
    ! xxx      Evaluation point
    ! x()      Data locations
    ! y()      Data values
    ! deg      Degree of interpolating polynomial
    ! Nv     Number of values in x() and in y()
    ! The starting index is assumed to be zero

    !if Pdeg is odd ...
    if pdeg<>2*(int(pdeg/2)) then&
&          shift=int((pdeg-1)/2)
    !if Pdeg is even ...
    if pdeg=2*(int(pdeg/2)) then&
&          shift=int(pdeg/2)

    ! Search for appropriate interval
    for isr=0 to Nv-1    ! search for interval
       if (x(isr)-xxx)*(x(isr+1)-xxx)<=0 then
          ! near beginning of table
          p=max(isr-shift,0)
          exit for
       end if
    next isr
    ! near end of table
    q=min(p+Pdeg,Nv)
    p=q-Pdeg
```

```
    ! Construct Newton Divided Difference
    ! Table on the interval
    mat ndd=zer
    for i=1 to q-p+1
        ndd(i,1)=y(p-1+i)
    next i
    for j=2 to q-p+1
        for i=1 to q-p+1-j+1
            ndd(i,j)=(ndd(i+1,j-1)-ndd(i,j-1))&
&                   /(x(p+i+j-2)-x(p+i-1))
        next i
    next j
end sub

sub newt     ! Evaluate the Newton Polynomial
    xtm=1
    ff(0)=ndd(1,1)
    for ord=1 to q-p
        xtm=xtm*(xxx-x(p+ord-1))
        ff(ord)=ff(ord-1)+ndd(1,ord+1)*xtm
    next ord
    fval=ff(q-p)
end sub

sub dataset     ! Generate the data
    Nv=25
    alpha=-.2   ! Parameter
    for i=0 to Nv
        ! Evenly or unevenly spaced data points
        x(i)=i-0.2*z*(-1)^i
        y(i)=25*exp(alpha*x(i))*cos(2*pi*x(i)/10)
        ! Function value at data points
    next i
end sub

end
```

In order to use the difference table for interpolation, after the coefficients are determined the polynomial must be evaluated. It is most efficient to do this using an implementation of the expression

$$P(x) = \ldots \{c_{n-2} + (x - x_{q-1}) \left[c_{n-1} + c_n (x - x_q) \right] \}$$

in a loop that evaluates the innermost terms first, and then proceeds with multiplication of each new $(x - x_i)$ factor and the addition of the c-term, repeating as many times as required. This is done in **subroutine newt** in the listing above.

Centered Difference Approximations

As mentioned above, Newton's divided difference method is awkward for estimating derivatives when the data is not evenly spaced. When the data is evenly spaced, however, some interesting results are obtained. Rather than approach them through Newton's divided differences, the same results will be developed here by means of Taylor series expansions. The resulting derivative approximations are called *centered* divided difference approximations. They are *centered* in that an equal number of data points are used on both sides of the evaluation point. They are *divided difference* approximations by virtue of their relation to Newton's divided difference table. The development is given for a particular situation using five points and then a computer program for generalization is presented.

Assuming that the data are evenly spaced at a step size h, consider the process of representing a function and as many of its derivatives as possible with N_p data points using Taylor series. The value N_p must be odd for a centered difference to result. Without any further loss of generality, the data may be considered as centered on a point with index 0 so that the data are $(x_0 - 2h, y_{-2})$, $(x_0 - h, y_{-1})$, (x_0, y_0), $(x_0 + h, y_1)$ and $(x_0 + 2h, y_2)$, emphasizing the even spacing. A Taylor series about x_0 will be made to fit each data value, as far as the derivative, dropping the remainder terms:

$$
\begin{aligned}
y_2 &= y_0 + \tfrac{2h}{1!} f'(x_0) + \tfrac{(2h)^2}{2!} f''(x_0) + \tfrac{(2h)^3}{3!} f'''(x_0) + \tfrac{(2h)^4}{4!} f''''(x_0) + \ldots \\
y_1 &= y_0 + \tfrac{h}{1!} f'(x_0) + \tfrac{h^2}{2!} f''(x_0) + \tfrac{h^3}{3!} f'''(x_0) + \tfrac{h^4}{4!} f''''(x_0) + \ldots \\
y_0 &= y_0 \\
y_{-1} &= y_0 - \tfrac{h}{1!} f'(x_0) + \tfrac{h^2}{2!} f''(x_0) - \tfrac{h^3}{3!} f'''(x_0) + \tfrac{h^4}{4!} f''''(x_0) + \ldots \\
y_{-2} &= y_0 - \tfrac{2h}{1!} f'(x_0) + \tfrac{(2h)^2}{2!} f''(x_0) - \tfrac{(2h)^3}{3!} f'''(x_0) + \tfrac{(2h)^4}{4!} f''''(x_0) + \ldots
\end{aligned}
$$

We will adopt the position that this is simply a system of linear equations to be solved for the function value, y_0, and the derivatives f', f'', f''', and f'''', in terms of the five data values, y_2, y_1, y_0, y_{-1}, and y_{-2}. For that purpose, the system is rewritten in matrix form as follows:

$$
\underbrace{\begin{bmatrix}
1 & \tfrac{2}{1!} & \tfrac{2^2}{2!} & \tfrac{2^3}{3!} & \tfrac{2^4}{4!} \\
1 & \tfrac{1}{1!} & \tfrac{1^2}{2!} & \tfrac{1^3}{3!} & \tfrac{1^4}{4!} \\
1 & 0 & 0 & 0 & 0 \\
1 & \tfrac{-1}{1!} & \tfrac{(-1)^2}{2!} & \tfrac{(-1)^3}{3!} & \tfrac{(-1)^4}{4!} \\
1 & \tfrac{-2}{1!} & \tfrac{(-2)^2}{2!} & \tfrac{(-2)^3}{3!} & \tfrac{(-2)^4}{4!}
\end{bmatrix}}_{[C_{ij}]}
\left\{ \begin{array}{c} y_0 \\ hf' \\ h^2 f'' \\ h^3 f''' \\ h^4 f'''' \end{array} \right\}
= \left\{ \begin{array}{c} y_2 \\ y_1 \\ y_0 \\ y_{-1} \\ y_{-2} \end{array} \right\}
$$

The general term in the coefficient matrix is

$$
C_{ij} = \frac{(2 - i + 1)^{j-1}}{(j-1)!}
$$

and the 2 in the numerator is the upper limit on the index range $(-2, \; -1, \; 0, \; 1, \; 2)$ so that for a number of points other than five, the appropriate value can be readily determined.

Both sides of the equation are now multiplied by the largest factorial and then a standard linear systems solution is carried out to give

$$\left\{ h^k f^{(k)} \right\} = \frac{1}{(\text{Max Factorial})} \underbrace{\left[(\text{Max Factorial}) \left[C_{ij} \right]^{-1} \right]}_{\text{All integers}} \left\{ y_i \right\} \quad .$$

The coefficient matrix $\left[\hat{C} \right] = \left[(\text{Max Factorial}) \left[C_{ij} \right]^{-1} \right]$ now consists entirely of integers. Looking at a single row of the result,

$$h^k f^{(k)} = \frac{\hat{C}_{k1}\, y_2 + \hat{C}_{k2}\, y_1 + \hat{C}_{k3}\, y_0 + \hat{C}_{k4}\, y_{-1} + \hat{C}_{k5}\, y_{-2}}{\text{Max Factorial}}$$

One of the characteristics of the central difference formulae is that the terms in the center carry the most weight, and terms toward the ends carry less weight. The nonzero $\hat{C}_{kj}$ farthest from the center, denoted $\check{C}_k$, is a common factor of all the $\hat{C}_{kj}$ and can be divided out of the relation to give

$$f^{(k)} = \frac{\frac{\hat{C}_{k1}}{\check{C}_k} y_2 + \frac{\hat{C}_{k2}}{\check{C}_k} y_1 + \frac{\hat{C}_{k3}}{\check{C}_k} y_0 + \frac{\hat{C}_{k4}}{\check{C}_k} y_{-1} + \frac{\hat{C}_{k5}}{\check{C}_k} y_{-2}}{\left(\frac{\text{Max Factorial}}{\check{C}_k} \right) h^k}$$

The final result is an expression for $f^{(k)}$, the kth derivative evaluated at x_0, as a sum of integer multiples of the y_i, the entire sum divided by an integer multiple of h^k. The whole process is implemented in the computer program listing that follows.

Program to Determine Centered Difference
Expressions of Various Orders

```
!                    CentDiff.Tru
! This program generates the coefficients for
! for approximating derivatives based on
! central differences
option nolet
option base 1
dim c(0,0),ci(0,0),one(0)
blk$="##################"
blk$=blk$&blk$&blk$&blk$
clear
hd1$="Expressions for Central Difference"
hd2$="Approximations to Higher Order Derivatives"
hd3$="Based on Evenly Spaced Data"
print using blk$: hd1$
print using blk$: hd2$
print using blk$: hd3$
print
s1$="Enter number of points to be used,"
s2$=" an ODD number"
print s1$&s2$
```

```
input Np
mat redim c(Np,Np),ci(Np,Np),one(Np)
ui=int(Np/2)
li=-ui
print "Upper index limit = Ui = ";ui
print "Lower index limit = Li = ";Li
for row=1 to Np
    fac=1
    for col=1 to Np
        if col=1 then
            fac=fac
        else if col>1 then
            fac=(col-1)*fac
        end if
        c(row,col)=(ui-row+1)^(col-1)/fac
    next col
next row
mf=fac
mat ci=inv(c)
mat ci=(mf)*ci ! These should all be integers
for i=1 to Np  ! Force them to be integers
    for j=1 to Np
        ci(i,j)=round(ci(i,j))
    next j
next i
hd0$="Expressions for Approximating Derivatives"
hd1$="Sum of terms of the form:"
hd1$=hd1$&"       multiplier x f sub (index)"
hd3$="        index            multiplier"
img$="          ##            ######.##"
print #od: hd0$
print #od
for id=1 to Np
    if id=1 then lbl$="Function itself"
    if id=2 then lbl$="1st Derivative"
    if id=3 then lbl$="2nd Derivative"
    if id=4 then lbl$="3rd Derivative"
    if id>4 then lbl$=str$(id-1)&"th Derivative"
    for j=1 to Np
        if ci(id,j)<>0 then
            div=abs(ci(id,j))
            exit for
        end if
    next j
    divisor=mf/div
```

```
    for i=1 to Np
        ci(id,i)=ci(id,i)/div
    next i
    clear
    print #od
    print #od
    print #od: "For "&lbl$
    print #od: hd1$
    s$="All terms divided by "
    s$=s$&str$(divisor)&" h^"&str$(id-1)
    print #od: s$
    print #od
    print #od: hd3$
    print #od
    for fi=1 to Np
        print #od, using img$: ui-fi+1,ci(id,fi)
    next fi
    print #od
    print #od: "Press  ENTER  to continue"
    if od=0 then get key xxx
next id
end
```

If this program is used to determine, for example, the coefficients for the case developed above, the results are

$$y_0 = y_0$$
$$f'(x_0) \approx \frac{-y_2 + 8y_1 + 0y_0 - 8y_{-1} + y_{-2}}{12h}$$
$$f''(x_0) \approx \frac{-y_2 + 16y_1 - 30y_0 + 16y_{-1} - y_{-2}}{12h^2}$$
$$f'''(x_0) \approx \frac{y_2 - 2y_1 + 0y_0 + 2y_{-1} - y_{-2}}{2h^3}$$
$$f''''(x_0) \approx \frac{y_2 - 4y_1 + 6y_0 - 4y_{-1} + y_{-2}}{h^4}$$

Similar results for other numbers of data points are readily obtained from the computer program.

References

Chapra, S.C., and Canale, R.P., *Numerical Methods for Engineers*, 2nd ed., McGraw-Hill, 1988.

Hoffman, J.D., *Numerical Methods for Engineers and Scientists*, McGraw-Hill, 1992.

Nakamura, S., *Applied Numerical Methods with Software*, Prentice-Hall, 1991.

Chapter 7

Applications of Green's Theorem

Green's Theorem

One of the major theorems of applied mathematics regarding line integrals is that known as Green's Theorem. It is particularly useful for the numerical evaluation of certain double integrals, as will be shown in this chapter. The theorem states:

> Let D be a domain of the x-y plane and let Γ be a piece-wise smooth simple closed curve in D whose interior is also in D. Let $P(x,y)$ and $Q(x,y)$ be functions defined and continuous and having continuous first partial derivatives in D. Then
>
> $$\oint_{\Gamma} P\,dx + Q\,dy = \iint_{A} \left(\frac{\partial Q}{\partial x} - \frac{\partial P}{\partial y} \right)\,dx\,dy$$
>
> where A is the closed region bounded by Γ. The integration process proceeds around the boundary curve with positive area on the left, that is, in a counter clockwise sense. (adapted from Kaplan, p. 239; see also Courant, p. 360).

For the purposes below, this theorem will be used to convert double integrals into cyclic line integrals. To accomplish this requires finding appropriate functions $P(x,y)$ and $Q(x,y)$, such that the integrand of the double integral is correctly represented by the required combination of derivatives.

One of the computational problems faced frequently is the need to perform an integration over a irregularly shaped domain. Adequately describing the limits of integration can be a time-consuming process. This is especially true if the boundary curve is known only in tabular form, as is often the case in non-idealized geometries. Provided that the boundary of the domain of integration can be reasonably approximated by a sequence of straight line segments drawn between a set of node points on the boundary, the integration can often by reduced by means of Green's Theorem to a line integral that is fairly easily evaluated numerically.

Consider the boundary curve broken into n segments by n nodes. For the cyclic integral, the integration may be taken as n line integrals, comprising the complete path of integration, Γ:

$$\oint_{\Gamma} (\cdots) \, ds = \int_{x_1,y_1}^{x_2,y_2} (\cdots) \, ds + \int_{x_2,y_2}^{x_3,y_3} (\cdots) \, ds + \cdots + \int_{x_n,y_n}^{x_1,y_1} (\cdots) \, ds$$

For numerical evaluation of a cyclic integral, the boundary curve Γ is replaced by an irregular n-sided polygon defined by the n nodes, numbered in the sense of positive motion along Γ and approximating the boundary curve. Each straight line segment falls into one of three categories:

Case 1   a vertical line, $x = constant$
Case 2   a horizontal line, $y = constant$, or
Case 3   an inclined line, $y = s_i (x - x_i) + y_i$

To prepare for numerical evaluation of the cyclic integrals, the contribution of a boundary segment from (x_i, y_i) to (x_{i+1}, y_{i+1}) for each of these cases is required. These expressions are developed by substituting the forms appropriate to each of the three cases in the integral and evaluating the integral between the two end points for each case. The resulting expressions for the contribution of each increment can then be implemented in a subroutine for the numerical evaluation.

Area, Centroid, and Inertia for an Irregular Planar Area

The area, centroid coordinates, and area moments of inertia for an irregular planar area are properties defined by the following double integrals:

$$A = \iint_A 1 \, dx \, dy$$

$$X_C = \frac{1}{A} \iint_A x \, dx \, dy$$

$$Y_C = \frac{1}{A} \iint_A y \, dx \, dy$$

$$I_{xx} = \iint_A y^2 \, dx \, dy$$

$$I_{yy} = \iint_A x^2 \, dx \, dy$$

In order to apply the ideas presented above regarding Green's Theorem, it is necessary to consider the integrand of each of the double integrals as being in the form $\frac{\partial Q}{\partial x} - \frac{\partial P}{\partial y}$ and then devising appropriate functions $P(x,y)$ and $Q(x,y)$. There is not a unique choice for P and Q, but, in fact, there are many satisfactory choices. For this application, the choice is made to take $Q(x,y) \equiv 0$, and then chose an appropriate function for $P(x,y)$. This has the effect of making the $\oint Q \, dy = 0$, and thus eliminating that part of the calculation.

The fundamental approximation to be made is that the boundary curve, Γ, may be approximated by an n-sided, irregular polygon. For the polygonal approximation, there are basically three cases:

Case 0   a vertical line, $x = constant$

Case 1   a horizontal line, $y = constant$

Case 2   an inclined line, $y = s_i (x - x_i) + y_i$

The first of these cases (number 0) is of no further concern, because the integral $\int P\,dx$ is zero along such a line segment. For the other two cases, it will be necessary to consider them in detail for each of the properties to be determined. Note that, before using the expressions developed for Case 2, it is necessary to evaluate the slope, s_i:

$$s_i = \frac{y_{i+1} - y_i}{x_{i+1} - x_i}$$

for the inclined line segment between (x_i, y_i) and (x_{i+1}, y_{i+1}).

Area

For the area calculation, $P = -y$ is a satisfactory choice. With this choice, $\frac{\partial P}{\partial y} = -1$, and the double integral is

$$A = \iint 1 \, dx \, dy$$

$$= -\iint \frac{\partial P}{\partial y} \, dx \, dy$$

$$= \iint \left(\frac{\partial Q}{\partial x} - \frac{\partial P}{\partial y} \right) \, dx \, dy$$

$$= \oint P \, dx + Q \, dy$$

$$= \oint P \, dx$$

$$= -\oint y \, dx$$

The negative sign may appear out of place here, but recall that the boundary curve is traversed with positive area on the left. Thus, for an area bounded by $x = a$, the x-axis, $x = b$, and a curve $y = f(x) > 0$ for $a \leq x \leq b$, in order to enclose the area on the left, the integration may be considered to begin at $(a, 0)$, move to the right along the x-axis, up along the line $x = b$, right to left along the curve $y = f(x)$, and down along the line $x = a$. On three legs of the circuit, the integral is zero (either $y = 0$, or $dx = 0$) and the expression for the area comes down to

$$A = -\oint y \, dx = -\int_b^a y \, dx = \int_a^b y \, dx$$

This is of course the familiar expression for the area, and the previous form is seen to be compatible with the more familiar form for this situation.

For Case 1, the value of y is a constant and can be taken out of the integration. For Case 2, the integral can be evaluated along the inclined line segment. The results are, for the area,

Case 1 $\qquad \Delta A = -y_i \left(x_{i+1} - x_i \right)$

Case 2 $\qquad \Delta A = - \left[\frac{1}{2} s_i \left(x_{i+1}^2 - x_i^2 \right) + \left(y_i - s_i x_i \right) \left(x_{i+1} - x_i \right) \right]$

The two formulae above, giving ΔA for Cases 1 and 2, are referred to as *increments in A*, or more simply just as *increments*. These are the incremental contributions to the cyclic integral from the segment i to $i+1$. As noted previously, there is no increment when the line segment is vertical because of the way in which the functions $Q(x, y)$ was chosen.

Centroidal Coordinates

For the calculation of X_C, the horizontal coordinate of the centroid, a suitable choice is $P = -xy$, for which $\frac{\partial P}{\partial y} = -x$. For the calculation of Y_C, the vertical coordinate of the centroid, a suitable choice is $P = -\frac{1}{2}y^2$, for which $\frac{\partial P}{\partial y} = -y$. The further development parallels that above, and is presented below in a parallel format.

$$X_C = \frac{1}{A} \iint x \, dx \, dy \qquad\qquad Y_C = \frac{1}{A} \iint_A y \, dx \, dy$$

$$= \frac{1}{A} \iint \left(-\frac{\partial P}{\partial y} \right) dx \, dy \qquad\qquad = \frac{1}{A} \iint \left(-\frac{\partial P}{\partial y} \right) dx \, dy$$

$$= \frac{1}{A} \iint \left(\frac{\partial Q}{\partial x} - \frac{\partial P}{\partial y} \right) dx \, dy \qquad\qquad = \frac{1}{A} \iint \left(\frac{\partial Q}{\partial x} - \frac{\partial P}{\partial y} \right) dx \, dy$$

$$= \frac{1}{A} \oint \left(P \, dx + Q \, dy \right) \qquad\qquad = \frac{1}{A} \oint \left(P \, dx + Q \, dy \right)$$

$$= \frac{1}{A} \oint P \, dx \qquad\qquad = \frac{1}{A} \oint P \, dx$$

$$= -\frac{1}{A} \oint xy \, dx \qquad\qquad = -\frac{1}{2A} \oint y^2 \, dx$$

As before, the incremental contributions for Cases 1 and 2 must be determined for each of these integrals. The process has been illustrated above, and the results are as follow:

Case 1 $\quad \Delta X_C = \frac{-1}{2A} y_i \left(x_{i+1}^2 - x_i^2 \right)$

Case 2 $\quad \Delta X_C = \frac{-1}{A} \left[\frac{1}{3} s_i \left(x_{i+1}^3 - x_i^3 \right) + \frac{1}{2} \left(y_i - s_i x_i \right) \left(x_{i+1}^2 - x_i^2 \right) \right]$

Case 1 $\quad \Delta Y_C = \frac{-1}{2A} y_i^2 \left(x_{i+1} - x_i \right)$

Case 2 $\quad \Delta Y_C = \frac{-1}{2A} [\frac{1}{3} s_i^2 \left(x_{i+1}^3 - x_i^3 \right) + s_i \left(y_i - s_i x_i \right) \left(x_{i+1}^2 - x_i^2 \right)$
$\qquad\qquad + \left(y_i^2 + s_i^2 x_i^2 \right) \left(x_{i+1} - x_i \right)]$

Area Moments of Inertia

The integrals involved for the area moments of inertia are

$$ I_{xx} = \iint_A y^2 \, dx \, dy $$

$$ I_{yy} = \iint_A x^2 \, dx \, dy $$

Following the line of development given above, for the moment of inertia with respect to the x-axis, a suitable choice for P is $P = -\frac{1}{3} y^3$ with the partial derivative $\frac{\partial P}{\partial y} = -y^2$. When carried through to the final form, the expression for I_{xx} is

$$ I_{xx} = -\frac{1}{3} \oint y^3 \, dx $$

The incremental expressions for this integral are

Case 1 $\quad \Delta I_{xx} = -\frac{1}{3} y_i^3 \left(x_{i+1} - x_i \right)$

Case 2 $\quad \Delta I_{xx} = -\frac{1}{3} [\frac{1}{4} s_i^3 \left(x_{i+1}^4 - x_i^4 \right) + s_i^2 \left(y_i - s_i x_i \right) \left(x_{i+1}^3 - x_i^3 \right)$
$\qquad\qquad + \frac{3}{2} s_i \left(y_i - s_i x_i \right)^2 \left(x_{i+1}^2 - x_i^2 \right) + \left(y_i - s_i x_i \right)^3 \left(x_{i+1} - x_i \right)]$

In the determination of I_{yy}, an appropriate choice for P is $P = -x^2 y$ with the partial derivative $\frac{\partial P}{\partial y} = -x^2$. This leads to the expression for I_{yy} as

$$ I_{yy} = -\oint x^2 y \, dx $$

The incremental expressions for I_{yy} are

Case 1 $\quad \Delta I_{yy} = -\frac{1}{3} y_i \left(x_{i+1}^3 - x_i^3 \right)$

Case 2 $\quad \Delta I_{yy} = -\frac{1}{4} s_i \left(x_{i+1}^4 - x_i^4 \right) - \frac{1}{3} \left(y_i - s_i x_i \right) \left(x_{i+1}^3 - x_i^3 \right)$

The calculations described above have been implemented in a simple program, PlnArea.Tru, for which a listing follows below. Following the listing, there are two recommended test cases to verify the program and a few comments on the quality of the results.

Program Listing for PlnArea.Tru

```
!                    PlnArea.Tru
! Geometric properties for a planar area
! of arbitrary shape
option nolet
option base 1
dim x(0),y(0)
s1$="Nodes must enclose the area,"
s11$="= positive area on the left"
s2$="This is sometimes described as"
s22$=" counter clockwise"
s3$="Enter the total number of nodes to be used"
clear
print #0
print #0: "                    PlnArea.Tru"
print #0
print #0: s1$&s11$
print #0: s2$&s22$
print #0
print #0: s3$
input n
mat redim x(n),y(n)
print
print "Enter the node coordinates in pairs, xi,yi"
for i=1 to n
    print "x(";i;"), y(";i;") = ?"
    input x(i),y(i)
next i
call calcs
print
print "              Results"
print
print "          Area = ";area
print "          Xc   = ";xc
print "          Yc   = ";yc
print "          Ixx  = ";ixx
print "          Iyy  = ";iyy

sub calcs
    asum=0      ! initialize sum for area
    xcsum=0     ! initialize sum for Xc
    ycsum=0     ! initialize sum for Yc
```

69

```
    ixxsum=0     ! initialize sum for Ixx
    iyysum=0     ! initialize sum for Iyy
    x1=x(1)
    y1=y(1)
    for i=1 to n
        i2=i+1
        if i2<=n then      ! Usual case
            x2=x(i2)
            y2=y(i2)
        else if i2>n then ! Last side
            x2=x(1)
            y2=y(1)
        end if
        if x2=x1 then
            da=0      ! vertical edge
            dxc=0
            dyc=0
            dixx=0
            diyy=0
            exit if
        else if y2=y1 then
            call cas1 ! horizontal segment
        else if y2<>y1 then
            call cas2 ! sloping segment
        end if
        ! Add increments ...
        asum=asum+da
        xcsum=xcsum+dxc
        ycsum=ycsum+dyc
        ixxsum=ixxsum+dixx
        iyysum=iyysum+diyy
        x1=x2
        y1=y2
    next i
    ! Final evaluations ...
    area=asum
    xc=xcsum/area
    yc=ycsum/area
    ixx=ixxsum/3
    iyy=iyysum
end sub

sub cas1
    da=-y1*(x2-x1)
    dxc=-y1*(x2^2-x1^2)/2
```

```
      dyc=-y1^2*(x2-x1)/2
      dixx=-y1^3*(x2-x1)
      diyy=-y1*(x2^3-x1^3)/3
end sub

sub cas2
      s=(y2-y1)/(x2-x1)
      da=-(s/2*(x2^2-x1^2)+(y1-s*x1)*(x2-x1))
      dxc=-(s/3*(x2^3-x1^3)+.5*(y1-s*x1)*(x2^2-x1^2))
      dyc=(s^2/3)*(x2^3-x1^3)+s*(y1-s*x1)*(x2^2-x1^2)
      dyc=-.5*(dyc+(y1-s*x1)^2*(x2-x1))
      dixx=s^3/4*(x2^4-x1^4)+s^2*(y1-s*x1)*(x2^3-x1^3)
      dixx=dixx+3*s/2*(y1-s*x1)^2*(x2^2-x1^2)
      dixx=-(dixx+(y1-s*x1)^3*(x2-x1))
      diyy=-s/4*(x2^4-x1^4)-(y1-s*x1)*(x2^3-x1^3)/3
end sub

end
```

It is useful to have a few test cases to validate a program such as this after it has been coded for the computer. The two test cases below will exercise all aspects of the program, and they are sufficiently simple to allow an exact calculation.

Rectangle: vertices at $(2,1)$, $(4,1)$, $(4,3)$, and $(2,3)$;

Triangle: vertices at $(2,0)$, $(3,0)$, and $(2,3)$.

	Rectangle	Triangle
Area $=$	4.0	1.5
$X_C =$	3.0	2.333333
$Y_C =$	2.0	1.0
$I_{xx} =$	17.33333	2.25
$I_{yy} =$	37.33333	8.25

The validity of the results when this program is applied to a complicated area with curved boundaries is dependent on the degree of approximation achieved by the n-sided polygon for which the program is exact. If the polygon deviates significantly from the intended curved boundary, the results will not be very good. Dividing the curved boundary into a greater number of parts will improve the results, and, in the limit as infinitely many divisions are used, the program result approaches the exact result, although the execution time becomes unacceptable! If a part of the boundary is straight, the program computes the contribution of that part of the boundary exactly; there is no benefit to dividing a straight edge into several segments.

Area, Volume, and Inertial Properties for a Solid of Revolution

Another application of the use of cyclic integrals is found in the determination of the surface area, volume, and inertial properties of a solid of revolution. Each of these quantities is describable as an integration over the area of the section; such an integration can be converted into a cyclic integral on the boundary of the section by means of Green's Theorem.

Basic Definitions

Consider a solid of revolution defined by the z-axis and a plane figure in the $r - z$ plane bounded by the curve Γ. For this solid, the surface area, volume, and axial position of the centroid are

$$A_S = 2\pi \oint_\Gamma r\, ds \qquad \text{Surface Area}$$

$$V = 2\pi \iint_A r\, dr\, dz \qquad \text{Volume}$$

$$Z_C = \frac{2\pi}{V} \iint_A rz\, dr\, dz \qquad \begin{array}{c}\text{Axial Position}\\ \text{of Centroid}\end{array}$$

where ds is differential arc length along the curve Γ, and A is the total cross section area with differential element $dr\, dz$.

The body is assumed to be homogeneous with mass density ρ, so the total mass is simply

$$M = \rho V \qquad \text{Mass}$$

and the center of mass coincides with the centroid of the volume. The cylindrical polar coordinate system, r-θ-z, is a principal coordinate system for this body, and the two transverse moments of inertia are equal. There are, therefore, only two distinct mass moments of inertia to be determined:

$$I_{rr} = \pi\rho \iint_A r^3\, dr\, dz + 2\pi\rho \iint_A rz^2\, dr\, dz \qquad \text{Transverse MMOI}$$

$$I_{zz} = 2\pi\rho \iint_A r^3\, dr\, dz \qquad \text{Axial MMOI}$$

Note that while the integral for the surface area involves a cyclic integral on Γ, all of the other properties require a double integration over the section area.

Application of Green's Theorem

With the application of Green's Theorem, these double integrals may be rewritten as cyclic integrals on Γ. As mentioned in the previous application, the choice of functions $P(x,y)$ and $Q(x,y)$ is not unique, and this application has not been handled in quite the same manner as the previous application. There are four basic integrals required, and after these are evaluated the expressions above for surface area, volume, axial position of the centroid, and mass moments of inertia are easily evaluated. The four basic integrals are

$$M_r = \iint_A r \, dr \, dz = \oint_\Gamma \left(rz \, dr + r^2 \, dz \right)$$

$$M_{rz} = \iint_A rz \, dr \, dz = \oint_\Gamma \left(rz^2 \, dr + \tfrac{3}{2} r^2 z \, dz \right)$$

$$M_{rz^2} = \iint_A rz^2 \, dr \, dz = -\oint_\Gamma \left(rz^3 \, dr + r^2 z^2 \, dz \right)$$

$$M_{r^3} = \iint_A r^3 \, dr \, dz = \oint_\Gamma \left(3r^3 z \, dr + r^4 \, dz \right)$$

These integrals show that the properties of interest may be considered as determined by a contribution from each segment of the boundary curve Γ. In the evaluation of these integrals, it is necessary that the boundary be traversed in the counterclockwise sense, that is, with positive area lying on the left of the boundary curve.

Polygonal Approximation

As in the previous application, the boundary curve is approximated by an irregular n-sided polygon, and any particular side falls into one of three categories:

Case 1 a cylinder, $r = constant$
Case 2 a flat, annular disk, $z = constant$
Case 3 a conical surface, $z = s_i (r - r_i) + z_i$

Because of the manner in which Green's Theorem was applied, all three cases can contribute to the value of the integral.

The incremental contribution for each case is determined as in the previous application. It is necessary to compute a slope, s_i, before applying any of the Case 3 increment formulae. The several incremental expressions are summarized in the following table:

	Case 1 Cylindrical Section $r = constant$	Case 2 Annular Disk $z = constant$	Case 3 Conical Section $z = s_i\,(r - r_i) + z_i$
$s_i =$	N/A	N/A	$(z_{i+1} - z_i)\,/\,(r_{i+1} - r_i)$
$\Delta\left(\frac{A_s}{2\pi}\right) =$	$r_i\,\lvert z_{i+1} - z_i \rvert$	$\frac{1}{2}\lvert r_{i+1}^2 - r_i^2 \rvert$	$\frac{1}{2}\sqrt{1 + s_i^2}\,\lvert r_{i+1}^2 - r_i^2 \rvert$
$\Delta\,(M_r) =$	$r_i^2\,(z_{i+1} - z_i)$	$\frac{1}{2}z_i\,(r_{i+1}^2 - r_i^2)$	$\frac{2}{3}s_i\,(r_{i+1}^3 - r_i^3)$ $-\frac{1}{2}(s_i r_i - z_i)\,(r_{i+1}^2 - r_i^2)$
$\Delta\,(M_{rz}) =$	$\frac{3}{4}r_i^2\,(z_{i+1}^2 - z_i^2)$	$\frac{1}{2}z_i^2\,(r_{i+1}^2 - r_i^2)$	$\frac{5}{8}s_i^2\,(r_{i+1}^4 - r_i^4)$ $-\frac{7}{6}s_i\,(s_i r_i - z_i)\,(r_{i+1}^3 - r_i^3)$ $+\frac{1}{2}(s_i r_i - z_i)^2\,(r_{i+1}^2 - r_i^2)$
$\Delta\,(M_{rz^2}) =$	$-\frac{1}{3}r_i^2\,(z_{i+1}^3 - z_i^3)$	$-\frac{1}{2}z_i^3\,(r_{i+1}^2 - r_i^2)$	$-\frac{2}{5}s_i^3\,(r_{i+1}^5 - r_i^5)$ $+\frac{5}{4}s_i^2\,(s_i r_i - z_i)\,(r_{i+1}^4 - r_i^4)$ $-\frac{4}{3}s_i\,(s_i r_i - z_i)^2\,(r_{i+1}^3 - r_i^3)$ $+\frac{1}{2}(s_i r_i - z_i)^3\,(r_{i+1}^2 - r_i^2)$
$\Delta\,(M_{r^3}) =$	$r_i^4\,(z_{i+1} - z_i)$	$\frac{3}{4}z_i\,(r_{i+1}^4 - r_i^4)$	$\frac{4}{5}s_i\,(r_{i+1}^5 - r_i^5)$ $-\frac{3}{4}(s_i r_i - z_i)\,(r_{i+1}^4 - r_i^4)$

A brief statement of the expressions for the increments and an example problem has been published previously (Doughty, 1981).

The theory developed above can be readily implemented in computer code. After this is done, the most complicated cross sections become tractable in numerical form. A program incorporating these ideas is listed below followed by a few comments on the program and the results for an example calculation.

Program Listing for SolRev.Tru

```
!               SolRev.Tru
! Geometric properties for a Solid of Revolution
! with an arbitrary cross section
option nolet
option base 1
dim r(0),z(0),x(0),y(0)
s1$="    Nodes must enclose the section area"
s2$="    counter clockwise in the R-Z plane"
s3$="Enter the total number of nodes "
s4$="to be used to define the section"
clear
```

```
print #0: "                      SOLREV"
print #0
print #0: s1$
print #0: s2$
print #0: "    (Positive area on the left.)"
print #0
print #0: s3$&s4$
input n
mat redim r(n),z(n),x(n),y(n)
print #0: "Enter the material density."
input p
print #0
print #0: "Enter section node coordinates, (Ri, Zi)"
for i=1 to n
    print #0: "R(";i;"), Z(";i;") = ?"
    input r(i),z(i)
next i
call calcs
call table

sub calcs
    g1=0        ! sum for perimeter
    g2=0        ! sum 1st radial moment of boundary
    g3=0        ! sum for sect area
    g4=0        ! sum 1st radial moment of area
    g5=0        ! sum 1st axial moment of area
    g6=0        ! sum product moment of area
    g7=0        ! sum mr^3 - axial moment of inertia
    g8=0        ! sum mrz^2 - transverse moment of inertia
    for i=1 to n
        i2=i+1
        if i2>n then   i2=1
        r1=r(i)
        z1=z(i)
        r2=r(i2)
        z2=z(i2)
        d1=sqr((r2-r1)^2+(z2-z1)^2)
        if r1=r2 then
           call cas1
        elseif z1=z2 then
           call cas2
        else
           call cas3
        end if
        ! Form sums
```

```
          g1=g1+d1
          g2=g2+d2
          g3=g3+d3
          g4=g4+d4
          g5=g5+d5
          g6=g6+d6
          g7=g7+d7
          g8=g8+d8
     next i
end sub

sub cas1       ! R=const
    d2=r1*abs(z2-z1)
    d3=0.5*r1*(z2-z1)
    d4=r1^2*(z2-z1)
    d5=-0.5*r1*(z2^2-z1^2)
    d6=0.75*r1^2*(z2^2-z1^2)
    d7=r1^4*(z2-z1)
    d8=-r1^2*(z2^3-z1^3)/3
end sub

sub cas2       ! Z=const
    d2=0.5*abs(r2^2-r1^2)
    d3=-0.5*z1*(r2-r1)
    d4=0.5*z1*(r2^2-r1^2)
    d5=-z1^2*(r2-r1)
    d6=0.5*z1^2*(r2^2-r1^2)
    d7=0.75*z1*(r2^4-r1^4)
    d8=-0.5*z1^3*(r2^2-r1^2)
end sub

sub cas3       ! Z=C(R-Ri)+Zi
    c=(z2-z1)/(r2-r1)
    f=c*r1-z1
    d2=0.5*sqr(1+c^2)*abs(r2^2-r1^2)
    d3=0.5*f*(r2-r1)
    d4=2*c/3*(r2^3-r1^3)-0.5*f*(r2^2-r1^2)
    t1=-2/3*c^2*(r2^3-r1^3)
    t2=1.5*c*f*(r2^2-r1^2)
    t3=-f^2*(r2-r1)
    d5=t1+t2+t3
    d6=5/8*c^2*(r2^4-r1^4)-7/6*c*f*(r2^3-r1^3)
    d6=d6+0.5*f^2*(r2^2-r1^2)
    t1=0.8*c*(r2^5-r1^5)
    t2=-0.75*f*(r2^4-r1^4)
```

```
      d7=t1+t2
      t1=-0.4*c^3*(r2^5-r1^5)
      t2=1.25*c^2*f*(r2^4-r1^4)
      t3=-4/3*c*f^2*(r2^3-r1^3)
      t4=0.5*f^3*(r2^2-r1^2)
      d8=t1+t2+t3+t4
end sub

sub table
      hdr$=        "      ###          ####.####      +####.####"
      clear
      print #0: "Print the data?     y/n"
      input g$
      if g$="y" then
         print #od: "            input Data"
         print #od: "      i              R(i)            Z(i)"
         for i=1 to n
             print #od, using hdr$: i,r(i),z(i)
         next i
         print #od
         print "Strike any key to continue"
         get key xxx
         clear
      end if
      print #od: "         Section Area Properties"
      print #od: "Perimeter = ";g1
      print #od: "Perimeter 1st moment = ";g2
      print #od: "Cross section area = ";g3
      print #od: "Section area radial 1st moment = ";g4
      print #od: "Section area axial 1st moment = ";g5
      print #od: "Centroid radius = ";g4/g3
      print #od: "Centroid axial position = ";g5/g3
      print #od
      print #od: "        Solid of Revolution Properties"
      print #od: "Surface area = ";2*pi*g2
      print #od: "Volume = ";2*pi*g4
      print #od: "Mass = ";2*pi*g4*p
      print #od: "Axial position of body centroid = ";g6/g4
      print #od: "   Principal Mass Moments of Inertia"
      print #od: "Irr = ";p*pi*(2*g8+g7)
      print #od: "Izz = ";2*pi*p*g7
      set cursor "off"
      get key xxx
end sub
end
```

The program begins with a data entry portion, and then the main calculation is done in a subroutine named `calcs` and subroutines that it calls, `cas1`, `cas2`, and `cas3`. These last three implement the detailed formulae given before the program listing for each of the three cases. The determination of which case applies is made in `calcs`, using a sequence of `if`-statements. The very last calculations and formatting of the printed results is done in the subroutine `table`. It is recommended that the user add plotting of the data after entering the node coordinates (r_i, z_i) that define the cross section. This will facilitate checking the data to assure that the correct section has been calculated.

Sample Calculation

Consider the triangular section defined by the following nodes:

Node i	r_i	z_i
1	1.00	0.50
2	3.00	0.00
3	2.00	2.00

When this section is revolved about the z-axis, the result is a ring of triangular section, having an edge on the inside, the outside, and the top. The mass density is taken as 1.0 for this example.

The computed results for this body of revolution are as follows:

Section Area Properties

Perimeter	$= 6.1003964$
Perimeter 1st Moment	$= 12.417439$
Cross Section Area	$= 1.75$
Section Area Radial 1st Moment	$= 3.50$
Section Area Axial 1st Moment	$= 1.4583333$
Centroid Radius	$= 2.00$
Centroid Axial Position	$= 0.8333333$

Solid of Revolution Properties

Surface Area	$= 78.02107$
Volume	$= 21.991149$
Mass	$= 21.991149$
Axial Position of Body Centroid	$= 0.8125$

Principal Mass Moment of Inertia

$$I_{rr} = 68.172561$$
$$I_{zz} = 98.960169$$

While only a simple cross section (a triangle) is involved here, the problem quickly becomes sufficiently complicated and quite laborious by manual methods. More complicated cross sections simply make the situation worse.

References

Courant, R. *Differential and Integral Calculus*, Vol. II, John Wiley & Sons, 1936,.

Doughty, S., "Calculating Properties for Solids of Revolution," *Machine Design*, 10 Dec., 1981, pp. 184 – 186.

Kaplan, W., *Advanced Calculus*, Addison-Wesley, 1952.

Chapter 8

Roots of a Single Equation

The problem of determining the roots of an equation occurs frequently, including:

- Determining the point of intersection of two curves, $f(x)$ and $g(x)$, by solving $f(x) = g(x)$

- Determining the location of a local maximum or minimum by solving the equation $f'(x) = 0$

- Determining the roots of the frequency equation in order to evaluate the natural frequencies of a vibrating beam

- Determining the roots of a polynomial for use in a partial fraction expansion

The two big categories to be considered here are (1) *general algebraic* and *transcendental equations*, and (2) *polynomial equations*. The literature of numerical analysis contains a great variety of methods for these two classes of problems. The presentation here will be confined to one reasonably reliable method for each class, with some comments regarding possible variations.

Numerical root finding is an iterative process directed toward the progressive refinement from a starting estimate of the root to a satisfactory approximation of the true root. Only in exceptional circumstances is the true root actually determined. Note the key terms in this statement: *starting estimate* and *satisfactory approximation*.

The starting estimate is just what it says, a beginning point. All numerical root solvers require some starting estimate, which for some methods and problems may be quite crude, while other situations require a rather good starting estimate, a starting point quite close to the root, if the process is to succeed.

If the process is successful, the starting estimate is progressively refined until a satisfactory estimate is obtained. The satisfactory estimate is "close enough," based on whatever criteria are relevant to the problem. The determination of the appropriate termination criteria is an important part of the problem definition, and realism demands that we understand in advance that the exact root cannot be determined in most cases. The successive estimates form a convergent sequence, which in the limit, is the true root. If the process is not successful, the sequence is not convergent.

General Algebraic and Transcendental Equations

The typical root finding problem for a non-polynomial algebraic equation or for a transcendental equation is limited to finding real roots only, and that limitation will be assumed in the discussion below. (Complex roots do exist in some cases, but their determination is beyond the scope of this presentation, except for polynomial equations to be considered later.) A wide variety of methods have been devised for problem of this type, but the most effective method for the whole class is Newton's method, which will be presented below.

Starting Estimates

For present purpose, the functions discussed will be continuous. An exception must be made when dealing with functions such as the tangent function for which there are well known discontinuities. These case can be handled as well, although they often require special care.

A root is a place where the function of interest has a zero value. Consider searching along the x-axis, examining the values of the function $f(x)$ at each point. For a continuous function, a change in algebraic sign for the function value indicates that a root has been passed. (Note that this is not the case for the tangent function when the sign changes in passing an odd multiple of $\frac{\pi}{2}$.) Thus searching the x-axis, keeping track of the function's sign, is one way to establish intervals containing roots, provided that the search step is sufficiently fine. If too large a step is made, it is possible to step over two roots, so that there is no apparent sign change, even though two sign changes have actually occurred. Another possibility with a large step is to pass three roots, recognizing only that one root has been passed. Although there are numerous perils associated with a root search, this is probably the most effective means of localizing each root *after* we have looked at the source of the problem to determine at the very least what part of the axis must be searched.

A search of the x-axis, in sufficiently small steps and over the appropriate interval, brackets each root in an interval such as $[x_a, x_b]$, where $f(x_a)$ and $f(x_b)$ are of opposite signs. This is the point of departure for many of the various methods, but Newton's method, that will be presented in detail below, requires a single starting estimate for the root. The value $\frac{x_a + x_b}{2}$ may be an appropriate starting estimate for Newton's method, although any information indicating that the root is closer to one end of the interval than the other should certainly be used. From this point on, it is assumed that a starting estimate is available for each root to be determined.

Newton's Method

Rather than attempt complete generality, the development of Newton's method will be done in the context of a typical problem. A more general statement will follow the specific application.

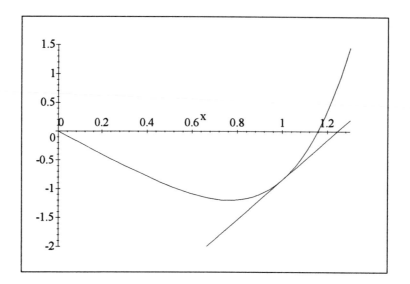

Figure 1: Plot of $f(x) = x^5 - x - \sin x$ and its tangent at $x = 1.0$.

Newton's Method by Example

Consider the problem of finding the roots of the following function:

$$f = x^5 - x - \sin x$$

This function has several roots, but let us suppose that we are interested in the root at approximately 1.2 as seen in Figure 1. This can serve as our starting estimate, but in order to better show how the process works, take the starting estimate as $x_{(1)} = 1$, where the subscript in parentheses is the revision number for the estimate. The value of $f\left(x_{(1)}\right)$ is $-.8414709848$. The term "residual" is often applied to the value of the function at the current estimate of the root, suggesting a quantity that we want to drive to zero but which at this point still remains nonzero. The solution process is essentially one of reducing the magnitude of the residual. For use in the next step, the slope at this point is obtained by differentiation and evaluation of the derivative at $x = 1$, $f'(1) = 3.459697694$.

The tangent line to the curve at this point is constructed as shown in Figure 1. Mathematically, this is the line

$$
\begin{aligned}
y(x) &= f'\left(x_{(1)}\right)\left[x - x_{(1)}\right] + y_{(1)} \\
y(x) &= 3.459697694\,(x - 1) - .8414709848 \\
&= 3.459697694\,x - 4.301168679
\end{aligned}
$$

The tangent line cuts the x-axis at $x_{(2)} = 1.243220957$, and from the graph it is evident that this is a better approximation to the root than the starting estimate, $x_{(1)} = 1$. The slope of the original function, $f'(x)$ evaluated at $x_{(2)}$ is 10.62262299.

The process is

- At each new estimate, $x_{(i)}$, evaluate the function $f\left(x_{(i)}\right)$ and its derivative, $f'\left(x_{(i)}\right)$.

- Using these numbers, determine the equation for the tangent line at the new estimate.

- Determine where that tangent line cuts the x-axis; this is the new estimate of the root.

- Repeat as many times as necessary.

For the example problem considered here, the final result is $x = 1.156904554$. If the process had been begun with our original starting estimate, $x_{(1)} = 1.2$, the result would not differ significantly (although there might be small differences in the last digit or two), but it would have been achieved in fewer iterations.

If, however, the original estimate were taken as $x_{(1)} = 0.6$, the slope would have been negative at that location, and the process would have converged to the root at zero. This is an equally valid root for the equation, but it is not the one of interest. There are other cases where a poor starting estimate can result in (1) very slow convergence, or (2) divergence, a situation where the estimate continues to change significantly and is not approaching any actual root of the equation.

The Algorithm

Consider the process again, this time expressed entirely in symbols. Suppose a current estimate, $x_{(i)}$ exists, where $i = 1$ denotes the starting estimate, or i has some higher value if the process has been underway previously. The function is evaluated at this estimate, $f\left(x_{(i)}\right)$, and also the derivative at that point, $f'\left(x_{(i)}\right)$. The equation of the line tangent to the curve at the current estimate is described by

$$f'\left(x_{(i)}\right) = \frac{y\left(x\right) - f\left(x_{(i)}\right)}{x - x_{(i)}}$$

or

$$y\left(x\right) = x f'\left(x_{(i)}\right) - x_{(i)} f'\left(x_{(i)}\right) + f\left(x_{(i)}\right)$$

The new estimate is the value of x where the tangent line intersects the x-axis, so that

$$y\left(x_{(i+1)}\right) = 0 = x_{(i+1)} f'\left(x_{(i)}\right) - x_{(i)} f'\left(x_{(i)}\right) + f\left(x_{(i)}\right)$$

which can be solved for the new estimate

$$x_{(i+1)} = x_{(i)} - \frac{f\left(x_{(i)}\right)}{f'\left(x_{(i)}\right)}$$

One area of difficulty is the appearance of the derivative in the denominator of the fraction. First, it is necessary to evaluate the derivative, which is in addition to the effort required to evaluate the function. More critically, if $\left| f'\left(x_{(i)}\right) \right|$ is small, then the estimate will make a large change. If that change is too large, the process may be shifted entirely away from the desired root. For this reason, it is sometimes desirable to arbitrarily reduce the adjustment made at each step by introducing a factor c, $0 < c < 1$, and using the expression

$$x_{(i+1)} = x_{(i)} - c \frac{f\left(x_{(i)}\right)}{f'\left(x_{(i)}\right)}$$

By appropriate selection of c the effect of a small slope can be minimized. It should be mentioned, however, that this can result in very slow convergence. The rationale is, however, that slow convergence is superior to no convergence.

If the slope is in fact zero at some point in the iteration, this can often be overcome by selecting a different starting estimate. It is usually a good idea, before plunging ahead with the new starting estimate, to carefully examine the graph of the function and to visualize the solution process.

The need to evaluate the derivative is a problem in some cases, particularly if the problem is difficult to express, such as finding the roots (x-values) of $f(p, q, r) = 0$, $p = p(x)$, $q = q(x)$, and $r = r(x)$. If the analytic expression of the derivative is difficult, it is often possible to numerically approximate the derivative as

$$f'\left(x_{(i)}\right) \approx \frac{f\left(x_{(i)} + \Delta x\right) - f\left(x_{(i)}\right)}{\Delta x}$$

where Δx is suitably chosen. When this is done, the algorithm is known as the secant method. Although this is often described as a separate, distinct method, there is much advantage to viewing this as simply a variation on Newton's method.

When to Quit

Throughout the above discussion, the process has been described at continuing until a satisfactory estimate has been achieved or until divergence is evident. Just how do we know when either of these has happened?

A satisfactory solution is defined by the situation that led to the problem at hand. If the purpose of this solution is to determine the equilibrium position of a mechanical system, where all of the dimensions are measured to a few thousandths of an inch, then we should probably try to obtain the equilibrium position to a similar accuracy. If, however, the problem is one on an astronomical scale, where the distances are rounded to millions of miles, then to seek a solution to within 500 miles would probably be inappropriate. The motivating situation clearly has to tell us what is a satisfactory solution.

Suppose that we have decided that we need to approximate the solution such that the magnitude of the final residual, the magnitude of the function value at our final estimate, is less than ε_1. Then after each step, we should test to see if

$$\left| f\left(x_{(i)}\right) \right| < \varepsilon_1 \quad ?$$

If the inequality is satisfied, then the iteration should be stopped; a satisfactory approximation has been determined. If the inequality is not satisfied, the iteration should continue. This is the first termination criterion.

There are situations where this criterion will never be satisfied. If the magnitude of the computed adjustment, $\left| \dfrac{f(x_{(i)})}{f'(x_{(i)})} \right|$, is less than the number of digits being carried in the estimate, the adjustment does not change the estimated value within the number of digits being carried. In this situation, the estimates sequence will seem to "stall," it will not change perceptibly, and yet it does not clearly diverge. This situation happens far too frequently to be ignored, so it is necessary to build-in a provision for stopping the iteration when this happens. The test to be made, again after each iteration, is

$$\left| \frac{f\left(x_{(i)}\right)}{f'\left(x_{(i)}\right)} \right| < \varepsilon_2 \quad ?$$

where ε_2 is based on the number of digits being carried, and perhaps other considerations as well. This becomes a second termination criterion.

The third circumstance in which the process should be stopped is divergence. Sometime the termination is made by the computer due to an overflow condition, but, in any event, the iteration loop should be limited to a reasonable number of iterations. This is easily accomplished in FORTRAN by a DO-loop, or in BASIC by a FOR-NEXT loop.

As a general rule, all three of these termination conditions should be included in the program. It is usually desirable to know which one of these conditions actually makes the termination, and that should be printed or displayed on the screen along with the numerical results. This strongly affects the confidence to be placed in the solution.

Example: Natural Frequencies of a Uniform Beam on Spring Supports

Consider a uniform beam (constant cross section and constant material properties) of length L and supported on two springs, one at $x = 0$ with stiffness k_1, and the second at $x = L$ with stiffness k_2. The motion of the beam is perpendicular to the length of the beam, and as it vibrates, (1) both springs deform and (2) the beam itself bends. The natural frequencies of vibration are determined from the requirement that the frequency determinant must be zero,

$$\begin{vmatrix} \cos \lambda L - \cosh \lambda L + 2\bar{k}_1 \sinh \lambda L & \sin \lambda L - \sinh \lambda L \\[2ex] \begin{aligned} -\sin \lambda L - \sinh \lambda L + \bar{k}_2 \left(\cos \lambda L + \cosh \lambda L \right) \\ + 2\bar{k}_1 \left(\cosh \lambda L - \bar{k}_2 \sinh \lambda L \right) \end{aligned} & \begin{aligned} -\cosh \lambda L + \cos \lambda L \\ + \bar{k}_2 \left(\sin \lambda L + \sinh \lambda L \right) \end{aligned} \end{vmatrix} = 0$$

where

$$\bar{k}_1 = k_1 / \left(EI\lambda^3 \right)$$

$$\bar{k}_2 = k_2 / \left(EI\lambda^3 \right)$$

E = Young's modulus for the beam material

I = area moment of inertia for the beam cross section

λ = separation constant from the governing partial differential equation, to be determined

After the separation constant is determined, the natural frequency of vibration is determined by

$$\omega = \lambda^2 \sqrt{\frac{EI}{\rho A}}$$

where

ρ = mass density of the beam material

A = cross sectional area of the beam

At first glance, this appears to be a formidable problem, but the fact is that the numerical aspect yields easily to Newton's method (the formulation is a bit more of a challenge!). The function for which roots are required is the value of the determinant given above, and this can be expanded to give

$$
\begin{aligned}
f(\lambda) \quad = \quad & -2\cos \lambda L \cosh \lambda L + \cos^2 \lambda L + 2\left(\cos \lambda L\right) k_2 \sinh \lambda L \\
& + \cosh^2 \lambda L - 2\left(\cosh \lambda L\right) k_2 \sin \lambda L + 2k_1 \sinh \lambda L \cos \lambda L \\
& + 4k_1 \left(\sinh \lambda L\right) k_2 \sin \lambda L + \sin^2 \lambda L - 2\left(\sin \lambda L\right) k_1 \cosh \lambda L \\
& - \sinh^2 \lambda L
\end{aligned}
$$

Newton's method requires the derivative of the function. That derivative is

$$
\begin{aligned}
\frac{df}{d\lambda} \quad = \quad & 2\left(\sin \lambda L\right) L \cosh \lambda L - 2\left(\cos \lambda L \sinh \lambda L\right) L \\
& -4\left(\sin \lambda L\right) L k_2 \sinh \lambda L - 4k_1 \left(\sinh \lambda L \sin \lambda L\right) L \\
& +4k_1 \left(\cosh \lambda L\right) L k_2 \sin \lambda L + 4k_1 \left(\sinh \lambda L\right) k_2 \left(\cos \lambda L\right) L
\end{aligned}
$$

A previous search has indicated that the first four roots are approximately at

Case	Starting Estimate
1	$\lambda_{1(1)} = 0.09$
2	$\lambda_{2(1)} = 0.12$
3	$\lambda_{3(1)} = 0.24$
4	$\lambda_{4(1)} = 0.39$

A program to refine these first four roots using Newton's method is given in the following example code. Note that the specific parameter values used are listed near the beginning of the code.

Example Program Using Newton's Method for Roots

```
option nolet
clear

!Problem Data ...
k1=75000                          ! Spring rate
k2=100000                         ! Spring rate
E=30000000                        ! Young's modulus
I=4.44                            ! Area moment of inertia
A=7.55                            ! Cross section area
L=20                              ! Length of beam
gamma=0.28                        ! Specific weight
grav=32.174*12                    ! Accel of gravity


! Note that  x  is equivalent to the variable  lambda
! as lambda was used in the analysis

! Loop for 4 cases ...
for icase=1 to 4
    if icase=1 then x=.09         ! Starting estimate, case 1
    if icase=2 then x=.12         ! Starting estimate, case 2
    if icase=3 then x=.24         ! Starting estimate, case 3
    if icase=4 then x=.39         ! Starting estimate, case 4
    flag=0                        ! Termination type marker

    c=0.5                         ! Step size control factor
    itmax=200                     ! Max number of iterations allowed

    ! Newton solution loop ...
    for iter=1 to itmax
        call eval                 ! Evaluate the function & deriv
        if abs(ff)<1e-8 then
           flag=1                 ! Termination for small residual
           exit for
        end if
        step=ff/fp
        x=x-c*step
        if abs(step)<1e-10 then
           flag=2                 ! Termination for small step
           exit for
        end if
    next iter
    omega=x^2*sqr(E*I*grav/(gamma*A))
    if flag=0 then print "Solution did not converge"
    if flag=1 then msg$="Solution terminated for small residual"
```

```
        if flag=2 then msg$="Solution terminated for small step"
     print "      omega    = ";omega
     print "      lambda*L = ";x*L
     print "      f(x)     = ";ff
     print "      ";msg$;"  iter     = ";iter
     print
next icase

sub eval
     ! This subroutine evaluates the function and its derivative
     ! at the current root estimate
     k1bar=k1/(E*I*x^3)
     k2bar=k2/(E*I*x^3)
     xL=x*L

     ff=-2*cos(xl)*cosh(xl)+(cos(xl))^2+2*k2bar*cos(xl)*sinh(xl)
     ff=ff+(cosh(xl))^2-2*k2bar*cosh(xl)*sin(xl)
     ff=ff+2*k1bar*sinh(xl)*cos(xl)
     ff=ff+4*k1bar*k2bar*sinh(xl)*sin(xl)+(sin(xl))^2
     ff=ff-2*k1bar*sin(xl)*cosh(xl)-(sinh(xl))^2

     fp=sin(xl)*cosh(xl)-cos(xl)*sinh(xl)-2*k2bar*sin(xl)*sinh(xl)
     fp=fp-2*k1bar*sinh(xl)*sin(xl)+2*k1bar*k2bar*cosh(xl)*sin(xl)
     fp=fp+2*k1bar*k2bar*sinh(xl)*cos(xl)
     fp=2*L*fp
end sub

end
```

The results produced by this program are summarized below:

	Case 1	Case 2	Case 3	Case 4
$\omega =$	1194.9366	2182.8765	9088.8356	24181.32
$\lambda L =$	1.7505763	2.366046	4.8279497	7.8749647
$f(\lambda) =$	$8.06 \cdot 10^{-9}$	$9.11 \cdot 10^{-9}$	$1.81 \cdot 10^{-7}$	$-4.00 \cdot 10^{-6}$
$iter =$	38	181	23	26
Termination:	small resid	small resid	small step	small step

It should be pointed out that this problem is one of those where it is necessary to decrease the step size to avoid divergence. The factor $c = 0.5$ is included for this purpose. If this factor is set to 1.0, the process diverges.

Another point needing comment is the iteration limit, itmax=200. This was necessary to obtain a fully converged solution for the second case. Early executions of the program started with a limit of 20, which was raised to 50, 70, 150, and finally 200. It certainly would not make sense to begin with the value 200.

Polynomial Equations

Polynomial equations are equations involving a polynomial expression equated to zero. In general, the polynomials may involve complex coefficients, but for the purpose of this discussion *only real coefficients will be considered*. Even with this assumption, the roots will often turn out to be complex. For the polynomial $P(x)$ having degree n,

$$P(x) = c_0 x^n + c_1 x^{n-1} + \ldots + c_{n-1} x + c_n$$

two important statements can be made:

- There are exactly n roots for $P(x)$;

- The roots are either real or they occur as complex conjugate pairs.

These two statements provide us with important guides in the root finding problem; they can be found in many algebra texts. They tell us (1) how many roots exist in all and (2) that complex roots will exist as conjugate pairs. Before going into the main development, two auxiliary matters are addressed first.

First, there is a second polynomial closely associated with the one given above and having exactly the same roots. Consider the polynomial above, evaluated at $1/x$:

$$
\begin{aligned}
P\left(\frac{1}{x}\right) &= c_0 \frac{1}{x^n} + c_1 \frac{1}{x^{n-1}} + \ldots + c_{n-2} \frac{1}{x^2} + c_{n-1} \frac{1}{x} + c_n \\
&= \frac{c_0 + c_1 x + \ldots + c_{n-2} x^{n-2} + c_{n-1} x^{n-1} + c_n x^n}{x^n} \\
&= \frac{1}{x^n} \left(c_0 + c_1 x + \ldots + c_{n-2} x^{n-2} + c_{n-1} x^{n-1} + c_n x^n \right) \\
&= \frac{1}{x^n} \tilde{P}(x)
\end{aligned}
$$

where

$$\tilde{P}(x) = c_0 + c_1 x + \ldots + c_{n-2} x^{n-2} + c_{n-1} x^{n-1} + c_n x^n$$

Thus the same list of coefficients, $(c_0, c_1, c_2, \ldots, c_{n-2}, c_{n-1}, c_n)$, is associated with both $P(x)$ and $\tilde{P}(x)$. Further, if $x = 1/r$ is a root of $P(x)$, then $x = r$ is a root of $\tilde{P}(x)$.

Secondly, there is the need for a means to evaluate the polynomial for complex arguments. The straight forward, brute-force approach promises a tremendous amount of complex arithmetic. Something better is needed. With some grouping and factoring, the polynomial can be written as

$$P(z) = c_0 z^n + c_1 z^{n-1} + \ldots + c_{n-1} z + c_n$$

89

$$
\begin{aligned}
P(z) &= \left(c_0 z^{n-1} + c_1 z^{n-2} + \ldots + c_{n-1}\right) z + c_n \\
&= \left[\left(c_0 z^{n-2} + c_1 z^{n-3} + \ldots + c_{n-2}\right) z + c_{n-1}\right] z + c_n \\
&\;\;\vdots \\
&= \left(\cdots \left\{\left[\left(c_0 z + c_1\right) z + c_2\right] z + c_3\right\} z \cdots\right) z + c_n
\end{aligned}
$$

where z is complex, $z = u + jv$, where u and v are real. The polynomial coefficients have been assumed to be real, but that does not mean that the value of the polynomial is real. In general, the value of the polynomial for a complex argument will be complex. The innermost term in the expression above is

$$
c_0 z + c_1 = \underbrace{c_0 u + c_1}_{\text{Real Part}} + j \underbrace{c_0 v}_{\text{Imag Part}}
$$

When this is multiplied by z and the next constant added, the result is

$$
\begin{aligned}
\left(c_0 z + c_1\right) z + c_2 &= \left(c_0 u + c_1 + j c_0 v\right)\left(u + jv\right) + c_2 \\
&= \left(c_0 u + c_1\right) u + j^2 \left(c_0 v\right) v + c_2 + j \left(c_1 v + 2 c_0 u v\right) \\
&= \underbrace{\left(c_0 u + c_1\right) u - c_0 v^2 + c_2}_{\text{Real Part}} + j \underbrace{\left(c_1 v + 2 c_0 u v\right)}_{\text{Imag Part}}
\end{aligned}
$$

Note that in each successive step, the previous complex value of the polynomial, (Real Part $+ j$Imag Part), is multiplied by the rules of complex arithmetic and the new constant is added to the resulting real part. The result is a new Real Part and an new Imag Part, to be carried to the next step until the process is completed. The final result is the complex value for the polynomial, and it may be useful to compute the magnitude of that complex result at this point. An example subroutine, PolyEval, that performs these steps is included later in this chapter, with the code implementing Bairstow's root finding algorithm.

Returning now to the main development, the polynomial $P(x)$ may be written in factored form as

$$
P(x) = c_n \left(x - z_1\right)\left(x - z_2\right)\left(\ldots\right)\left(x - z_n\right)
$$

provided that roots z_i, $i = 1, 2, \ldots n$ are known. (Typically, the roots are not known, and one of the reasons for wanting to find the roots is to be able to factor the polynomial.) This suggests that the polynomial could also be written in the form

$$
P(x) = \left(x^2 - ux - v\right)\left(d_{n-2} + d_{n-3}x + d_{n-4}x^2 + \ldots + d_0 x^{n-2}\right)
$$

provided that u and v are chosen appropriately. The scheme to be developed here is a method for evaluating u and v, and once they have been appropriately determined, the roots of the quadratic factor can be readily determined. After those two roots have been evaluated, the process continues by seeking roots for the reduced polynomial,

$$P_{\text{Reduced}}(x) = d_0 x^{n-2} + d_1 x^{n-3} + \ldots + d_{n-4} x^2 + d_{n-3} x + d_{n-2}$$

The advantages in looking for a quadratic factor as opposed to a linear factor are: (1) two roots are determined with each factorization, and (2) the amount of complex arithmetic is greatly reduced. The process of suitably adjusting u and v could be done by any of several methods including a two dimensional version of Newton's method, although this would involve a substantial amount of complex computation and be unpleasantly sensitive to the starting estimates. The procedure developed below is called Bairstow's method, and it uses a very clever approach to the evaluation of the necessary partial derivatives.

For arbitrary choices of u and v, the quadratic factor does not divide the polynomial exactly, and there is both a quotient polynomial and a remainder

$$
\begin{aligned}
P(x) &= \left(x^2 - ux - v\right)\left(d_0 x^{n-2} + d_1 x^{n-3} + \ldots + d_{n-3} x + d_{n-2}\right) + R(x) \\
R(x) &= r_1 x + r_0
\end{aligned}
$$

Multiplying out the factored form and comparing it to the original form shows that

Coef of	Relation
x^n	$d_0 = c_0$
x^{n-1}	$d_1 = c_1 + u d_0$
x^{n-2}	$d_2 = c_2 + u d_1 + v d_0$
x^{n-3}	$d_3 = c_3 + u d_2 + v d_1$
x^{n-4}	$d_4 = c_4 + u d_3 + v d_2$
$\vdots$	$\vdots$
x^2	$d_{n-2} = c_{n-2} + u d_{n-3} + v d_{n-4}$
x^1	$r_1 = c_{n-1} + u d_{n-2} + v d_{n-3}$
x^0	$r_0 = c_n + v d_{n-2}$

where the objective is to make $r_0 = r_1 = 0$, and thus make $\left(x^2 - ux - v\right)$ a factor of the original polynomial.

If the d-sequence is calculated, following the established pattern, to the last two elements, they are

$$
\begin{aligned}
d_{n-1} &= c_{n-1} + u d_{n-2} + v d_{n-3} \\
d_n &= c_n + u d_{n-1} + v d_{n-2}
\end{aligned}
$$

By inspection, it is evident that $r_1 = d_{n-1}$ and that $r_0 = d_n - u d_{n-1}$. Thus, if d_n and d_{n-1} are both reduced to zero, then r_0 and r_1 will be zero. This is the goal.

In the further development, it will be useful to have a means to evaluate $\frac{\partial d_{n-1}}{\partial u}$, $\frac{\partial d_{n-1}}{\partial v}$, $\frac{\partial d_n}{\partial u}$, and $\frac{\partial d_n}{\partial v}$. That evaluation will be addressed at this point, beginning by differentiating each of the d_i with respect to first u:

Derivative	Basic Expression	In terms of the e's	Def of e_i
$\frac{\partial d_0}{\partial u}$	$= 0$		
$\frac{\partial d_1}{\partial u}$	$= d_0$		$= e_0$
$\frac{\partial d_2}{\partial u}$	$= d_1 + u\frac{\partial d_1}{\partial u}$	$= d_1 + ue_0$	$= e_1$
$\frac{\partial d_3}{\partial u}$	$= d_2 + u\frac{\partial d_2}{\partial u} + v\frac{\partial d_1}{\partial u}$	$= d_2 + ue_1 + ve_0$	$= e_3$
$\vdots$			
$\frac{\partial d_{n-2}}{\partial u}$	$= d_{n-3} + u\frac{\partial d_{n-3}}{\partial u} + v\frac{\partial d_{n-4}}{\partial u}$	$= d_{n-3} + ue_{n-4} + ve_{n-5}$	$= e_{n-3}$
$\frac{\partial d_{n-1}}{\partial u}$	$= d_{n-2} + u\frac{\partial d_{n-2}}{\partial u} + v\frac{\partial d_{n-3}}{\partial u}$	$= d_{n-2} + ue_{n-3} + ve_{n-4}$	$= e_{n-2}$
$\frac{\partial d_n}{\partial u}$	$= d_{n-1} + u\frac{\partial d_{n-1}}{\partial u} + v\frac{\partial d_{n-2}}{\partial u}$	$= d_{n-1} + ue_{n-2} + ve_{n-3}$	$= e_{n-1}$

The table above shows that all of the partial derivatives with respect to u can be evaluated recursively. The e_i are simply shorter notations for these derivatives. The general expression for e_i is

$$e_i = d_i + ue_{i-1} + ve_{i-2}$$

where e_i is zero for negative subscripts. Now consider the second matter of the partial derivatives of the d's with respect to v:

Derivative	Basic Expression	In terms of the e's	As a single e
$\frac{\partial d_0}{\partial v}$	$= 0$		
$\frac{\partial d_1}{\partial v}$	$= 0$		
$\frac{\partial d_2}{\partial v}$	$= d_0 + u\frac{\partial d_1}{\partial v} = d_0$		$= e_0$
$\frac{\partial d_3}{\partial v}$	$= d_1 + u\frac{\partial d_2}{\partial v} + v\frac{\partial d_1}{\partial v}$	$= d_1 + ue_0$	$= e_1$
$\frac{\partial d_4}{\partial v}$	$= d_2 + u\frac{\partial d_3}{\partial v} + v\frac{\partial d_2}{\partial v}$	$= d_2 + ue_1 + ve_0$	$= e_2$
$\vdots$	$\vdots$	$\vdots$	$\vdots$
$\frac{\partial d_{n-1}}{\partial v}$	$= d_{n-3} + u\frac{\partial d_{n-2}}{\partial v} + v\frac{\partial d_{n-3}}{\partial v}$	$= d_{n-3} + ue_{n-4} + ve_{n-5}$	$= e_{n-3}$
$\frac{\partial d_n}{\partial v}$	$= d_{n-2} + u\frac{\partial d_{n-1}}{\partial v} + v\frac{\partial d_{n-2}}{\partial v}$	$= d_{n-2} + ue_{n-3} + ve_{n-4}$	$= e_{n-2}$

The e's were defined as derivatives of the d's with respect to u, but as shown above, they also represent, with a shift in index, the derivatives of the d's with respect to v. The partial derivatives that were originally sought are, therefore,

$$\frac{\partial d_{n-1}}{\partial u} = e_{n-2}$$
$$\frac{\partial d_{n-1}}{\partial v} = e_{n-3}$$
$$\frac{\partial d_n}{\partial u} = e_{n-1}$$
$$\frac{\partial d_n}{\partial v} = e_{n-2}$$

To search for the appropriate value of u and v, a process similar to Newton's method will be used, this time, however, in two unknowns. Consider d_{n-1} and d_n as functions of u and v, and consider a Taylor series expansion of both functions,

$$d_{n-1}\left(u_{(i)}+\Delta u, v_{(i)}+\Delta v\right) = d_{n-1}\left(u_{(i)}, v_{(i)}\right) + \frac{\partial d_{n-1}}{\partial u}\Delta u + \frac{\partial d_1}{\partial v}\Delta v + \dots$$
$$= d_{n-1}\left(u_{(i)}, v_{(i)}\right) + e_{n-2}\Delta u + e_{n-3}\Delta v + \dots$$

$$d_n\left(u_{(i)}+\Delta u, v_{(i)}+\Delta v\right) = d_n\left(u_{(i)}, v_{(i)}\right) + \frac{\partial d_n}{\partial u} + \Delta u + \frac{\partial d_n}{\partial v}\Delta v + \dots$$
$$= d_n\left(u_{(i)}, v_{(i)}\right) + e_{n-1}\Delta u + e_{n-2}\Delta v + \dots$$

Since the objective is to reduce both d_{n-1} and d_n to zero, suppose that the left sides of this equations are each zero, and solve for the changes required to accomplish that reduction,

$$u_{(i+1)} = u_{(i)} + \frac{e_{n-3}d_n - e_{n-2}d_{n-1}}{e_{n-2}^2 - e_{n-1}e_{n-3}}$$
$$v_{(i+1)} = v_{(i)} + \frac{e_{n-1}d_{n-1} - e_{n-2}d_n}{e_{n-2}^2 - e_{n-1}e_{n-3}}$$

As a brief summary, before looking at the code, the method consist of determining a quadratic factor of the form $x^2 - ux - v$ that will exactly divide the original polynomial. The variables to be adjusted to accomplish this are u and v, and their adjustment is along lines similar to Newton's method, with Bairstow's clever approach to the evaluation of the partial derivatives. After a pair of roots is found, the process is applied to the reduced polynomial. This is repeated until there is either a linear expression remaining or no remainder at all.

Polynomial Root Finding Example

Computer code for Bairstow's method is presented below in the context of an example problem. Bairstow's method is included in a subroutine `Bairstow`, which calls a second subroutine `Roots`. The program also includes a subroutine `PolyEval` for the evaluation of a polynomial for a complex argument. The problem considered for this example is the determination of the roots of the fourth degree polynomial

$$x^4 - 16 = 0$$

for which the roots are $2+j0$, $-2+j0$, $0+j2$, and $0-j2$. The data for the problem are included in two data statements in the main program, the first giving the degree of the polynomial and the second specifying the coefficients. The user can readily modify this code to deal with other polynomials by simply changing these two lines.

Example Code for Polynomial Roots by Bairstow's Method

```
! Polynomial roots by Bairstow's method
option nolet
option base 0
dim c(0),rc(0),e(0),co(0),rr(0),ir(0),resid(0),msg$(0)
clear
```

93

```
! Test Case:                    z^4 - 16 = 0
! List the value for N
data 4
! List the coefficients, beginning with highest order
data 1,0,0,0,-16     ! roots, 2+j0, -2+j0, 0+j2, 0-j2

read n
mat redim c(n)
for i=0 to n
    read c(i)
next i

call Bairstow(n,c(),rc(),e(),co(),rr(),ir(),resid(),msg$())

! Print the results
print "Root","Root","Magnitude of","Type of Root"
print "Real Part","Imag Part","Residual"
for i=1 to n
    print rr(i),ir(i),resid(i),msg$(i)
next i
end

sub Bairstow(n,c(),d(),e(),oc(),zr(),zi(),res(),lbl$())
    option base 0
    ! n       = degree of polynomial
    ! c()     = vector of polynomial coef
    ! d()     = vector of coef for reduced polynomial
    ! e()     = vector of partials derivatives
    ! oldn    = degree of original polynomial
    ! oc()    = vector of original polynomial coef
    ! zr()    = vector of root real parts
    ! zi()    = vector of root imag parts
    ! res()   = vector of residual magnitudes
    ! lbl$()  = vector of root type labels

    ! Save original degree & coefficients for later root checking
    mat oc=c                    ! Save original polynomial
    oldn=n                      ! Save original degree

    v=1                         ! Starting estimate for v
    u=1                         ! Starting estimate for u
    eps=1e-10                   ! Termination criterion
    itmax=1000                  ! Max number of iterations
    rc=0                        ! Count of roots found
```

```
mat redim oc(oldn),lbl$(oldn),zr(oldn),zi(oldn),res(oldn)
mat redim c(n),d(n),e(n)

do while n>=3

    erru=2*eps
    errv=2*eps
    for iter=1 to itmax
        if erru<=eps and errv<=eps then exit for
        for i= 0 to n
            if i=0 then
                di1=0
                di2=0
                ei1=0
                ei2=0
            else if i=1 then
                di1=d(0)
                di2=0
                ei1=e(0)
                ei2=0
            else
                di1=d(i-1)
                di2=d(i-2)
                ei1=e(i-1)
                ei2=e(i-2)
            end if
            d(i)=c(i)+u*di1+v*di2
            e(i)=d(i)+u*ei1+v*ei2
        next i
        determ=e(n-2)^2-e(n-1)*e(n-3)
        if determ<>0 then
            delu=(-e(n-2)*d(n-1)+e(n-3)*d(n))/determ
            delv=(e(n-1)*d(n-1)-e(n-2)*d(n))/determ
            u=u+delu
            v=v+delv
            if abs(u)>1e-2 then
                erru=abs(100*delu/u)
            else
                erru=delu*1000
            end if
            if abs(v)>1e-2 then
                errv=abs(100*delv/v)
            else
                errv=delv*1000
```

```
                end if
             else                         ! Det was zero, try again
                u=u+1
                v=v+1
                iter=0
             end if
          next iter
          ! Determine the roots of quadratic factor
          call roots(u,v,rr,ri,oc(),oldn,lbl$(),rc,zr(),zi(),res())
          n=n-2
          mat redim c(n)
          mat c=d
       loop
       if iter<itmax then              ! Deal with last 1 or 2 roots
          if n=2 then                  ! Quadratic polynomial remaining
             u=-c(1)/c(0)
             v=-c(2)/c(0)
             call roots(u,v,rr,ri,oc(),oldn,lbl$(),rc,zr(),zi(),res())
          else                         ! Linear polynomial remaining
             rr=-c(1)/c(0)
             ri=0
             call polyeval(oldn,oc(),rr,ri,preal,pimag,resid)
             rc=rc+1
             zr(rc)=rr
             zi(rc)=ri
             res(rc)=resid
             lbl$(rc)="Single Real Root"
          end if
       else
          print #0: "*******  Maxiumum Iterations Exceeded  *******"
       end if
       n=oldn                          ! Restore original value of n
    end sub

    sub roots(u,v,rr,ri,oc(),oldn,lbl$(),rc,zr(),zi(),res())
       disc=u^2+4*v
       if disc>0 then
          rr=(u-sqr(disc))/2            ! Distinct real roots
          ri=0
          call polyeval(oldn,oc(),rr,ri,preal,pimag,resid)
          rc=rc+1
          zr(rc)=rr
          zi(rc)=ri
          res(rc)=resid
          lbl$(rc)="Distinct Real Root-#1"
```

```
      rr=(u+sqr(disc))/2
      ri=0
      call polyeval(oldn,oc(),rr,ri,preal,pimag,resid)
      rc=rc+1
      zr(rc)=rr
      zi(rc)=ri
      res(rc)=resid
      lbl$(rc)="Distinct Real Root-#2"
   else if disc=0 then            ! Repeated real roots
      rr=u/2
      ri=0
      call polyeval(oldn,oc(),rr,ri,preal,pimag,resid)
      rc=rc+1
      zr(rc)=rr
      zi(rc)=ri
      res(rc)=resid
      lbl$(rc)="Repeated Real Root-#1"
      rc=rc+1
      zr(rc)=rr
      zi(rc)=ri
      res(rc)=resid
      lbl$(rc)="Repeated Real Root-#2"
   else if disc<0 then            ! Complex roots
      rr=u/2
      ri=sqr(-disc)/2
      call polyeval(oldn,oc(),rr,ri,preal,pimag,resid)
      rc=rc+1
      zr(rc)=rr
      zi(rc)=ri
      res(rc)=resid
      lbl$(rc)="Complex Conjugate Pair-#1"
      ri=-ri
      call polyeval(oldn,oc(),rr,ri,preal,pimag,resid)
      rc=rc+1
      zr(rc)=rr
      zi(rc)=ri
      res(rc)=resid
      lbl$(rc)="Complex Conjugate Pair-#2"
   end if
end sub

sub polyeval(n,c(),rr,ri,rp,ip,resid)
   ! Routine to evaluate polynomial for complex argument
   !   rr  - Real part of root
   !   ri  - Imag part of root
```

```
!   rp  - Real part of polynomial value
!   ip  - Imag part of polynomial value
rp=c(1)+c(0)*rr
ip=c(0)*ri
for i=2 to n
    zr=c(i)+rp*rr-ip*ri
    zi=rp*ri+ip*rr
    rp=zr
    ip=zi
next i
resid=sqr(rp^2+ip^2)
end sub
```

The subroutine Roots obtains the roots of the quadratic factor by use of the quadratic formula. An interesting alternative at this point, suggested by Conte for the situation where the discriminant is positive but not included in the code here, is to use the quadratic formula in a modified form, obtaining the smallest root by

$$x_{\text{Smaller Magnitude}} = \frac{-2v}{u + sgn\left(u\right)\sqrt{u^2 + 4v}}$$

where the $sgn\left(u\right)$ function assures that both terms will be of the same sign. This will avoid the problem of a difference in the denominator, resulting in a small divisor with round-off error and a potentially significant error in the quotient. The second root is then found by the fact that the negative of the coefficient of the linear term is the sum of the roots,

$$x_{\text{Larger Magnitude}} = -x_{\text{Smaller Magnitude}} + u$$

in the current notation.

References

Chapra, S.C., and Canale, R.P., *Numerical Methods for Engineers*, 2nd ed., McGraw-Hill, 1988.

Conte, S.D., *Elementary Numerical Analysis*, McGraw-Hill, 1965.

Jaluria, Y., *Computer Methods for Engineering*, Allyn and Bacon, 1988.

Chapter 9

Systems of Nonlinear Equations

Chapter 2 discussed the numerical solution of a system of linear equations and Chapter 8 considered the numerical solution of a single nonlinear equation. Ideas from both of these discussions are joined in this section to develop a method for the simultaneous, numerical solution of a system of nonlinear equations. This technique is known as the Newton - Raphson Method, and it is rather clearly an extension of Newton's Method (Chapter 8) to multiple dimensions.

Newton-Raphson Method

Statement of the Problem

The Newton-Raphson solution technique is an iterative method for solving a system of simultaneous, nonlinear algebraic equations. It has application to many problems of applied mathematics, physics and engineering. Consider the system of equations to be written in the form

$$\begin{aligned} f_1(x_1, x_2, \cdots x_n) &= 0 \\ f_2(x_1, x_2, \cdots x_n) &= 0 \\ &\vdots \\ f_n(x_1, x_2, \cdots x_n) &= 0 \end{aligned}$$

where this system is to be solved for $x_1, x_2, \cdots x_n$.

It is convenient to consider the estimates for the solutions as column vector (a one-dimensional array),

$$\{x\} = col(x_1, x_2, \cdots x_n)$$

and the values of the functions that make up the system of equations as another column vector,

$$\{f\} = col(f_1, f_2, \cdots f_n)$$

With these definitions, the problem may be written in a very compact form as

$$\{f(\{x\})\} = \{0\}$$

If a set of values are arbitrarily selected for $\{x\}$ and used to evaluate $\{f\}$, it is most unlikely that the result will be the zero vector, $\{0\}$. This is true, even if the values selected for $\{x\}$ are the best possible estimates available without *a priori* knowledge of the solution. The result of evaluating $\{f\}$, based on these best estimates, is called the *residual vector*, that which remains non-zero. The focus of the Newton-Raphson process is to make the residuals acceptably small. It is unlikely that an exact solution will ever be achieved ($\{f(\{x\})$ becoming *exactly* $\{0\}$), but it can be made as small as desired within the limitations of our ability to compute.

The Process

Let $\{x\}_{(i)}$ denote the ith estimate for the solution vector. The objective is to achieve an improved estimate, $\{x\}_{(i+1)}$, forming a convergent sequence of estimates. Consider a Taylor series expansion for $\{f\}$:

$$
\begin{aligned}
f_1(\{x\}_{(i+1)}) &= f_1(\{x\}_{(i)}) + \frac{\partial f_1}{\partial x_1}\Delta x_1 + \frac{\partial f_1}{\partial x_2}\Delta x_2 + \cdots \\
f_2(\{x\}_{(i+1)}) &= f_2(\{x\}_{(i)}) + \frac{\partial f_2}{\partial x_1}\Delta x_1 + \frac{\partial f_2}{\partial x_2}\Delta x_2 + \cdots \\
&\vdots \qquad\qquad \vdots
\end{aligned}
$$

This can be written compactly in matrix form as

$$
\{f(\{x\}_{(i+1)})\} = \{f(\{x\}_{(i)})\} +
\begin{bmatrix}
\frac{\partial f_1}{\partial x_1} & \frac{\partial f_1}{\partial x_2} & \frac{\partial f_1}{\partial x_3} & \cdots \\
\frac{\partial f_2}{\partial x_1} & \frac{\partial f_2}{\partial x_2} & \frac{\partial f_2}{\partial x_3} & \cdots \\
\frac{\partial f_3}{\partial x_1} & \frac{\partial f_3}{\partial x_2} & \frac{\partial f_3}{\partial x_3} & \cdots \\
\vdots & \vdots & \vdots & \ddots
\end{bmatrix}_{\{x\}_{(i)}}
\begin{Bmatrix}
\Delta x_1 \\
\Delta x_2 \\
\Delta x_3 \\
\vdots
\end{Bmatrix}
+ HOTs
$$

or, in a more compact form,

$$
\{f(\{x\}_{(i+1)})\} = \{f(\{x\}_{(i)})\} + \left[\frac{\partial f_i}{\partial x_j}\right]_{x_{(i)}} (\{x\}_{(i+1)} - \{x\}_{(i)}) + \text{Higher Order Terms}
$$

Notice that the matrix of partial derivatives is evaluated at $\{x\}_{(i)}$, the current estimate of the solution. This square matrix is quite important and is given a special name: the Jacobian matrix. In the following, this matrix will be denoted $[J]$. The "higher order terms," or HOTs, refers to products of the vector $\{\Delta x\}$ with itself, producing second and higher powers of $\Delta x_1, \Delta x_2, \cdots$. At this point, these higher order terms will be dropped, leaving a *linear approximation* to the nonlinear equations. The linear approximation is then solved by the standard methods of matrix algebra to give $\{\Delta x\}$, the *adjustment* to the solution estimate $\{x\}$.

$$
\{\Delta x\} = (\{x\}_{(i+1)} - \{x\}_{(i)}) = -\left[\frac{\partial f_i}{\partial x_j}\right]_{x_{(i)}}^{-1} \{f(\{x\}_{(i)})\}
$$

$$
\{x\}_{(i+1)} = \{x\}_{(i)} + \{\Delta x\}
$$

The entire process of linear approximation and solution of the linear approximation is repeated until *(1) the residuals are considered acceptably small*, or *(2) the solution does not move*. This is an *iterative* process, and it can be continued as long as progress is being made in approximating the solution. It is suggested that this be done under a finite loop (a FORTRAN DO-loop or a BASIC FOR-NEXT loop), to avoid the possibility of an infinite loop. The infinite loop would arise when the termination criteria are never met. If a finite loop is used, it is necessary to determine when the program moves past the iterative loop, whether it has actually found satisfactory estimates for the solution, or whether it moved ahead simply because the loop expired. Appropriate tests must be provided to make this determination, since disastrous results can arise from further work based on a "false solution," the solution estimates when the loop expired.

In the event that the solution has converged to a satisfactory approximation, the residuals will all be small, but how small is small? This is definitely a judgment matter, and the analyst must determine exact criterion to be used for each specific problem. For most cases, the solution is satisfactory if the residuals are four orders of magnitude less than the typical term in the equation; it may be desirable in some situations to set a stringent criterion, such as six orders of magnitude less than the typical term, or it may be necessary to relax the criterion to three orders of magnitude. It is certainly possible to develop a test to be applied to each residual individually that will determine when the established criterion is satisfied. At that point, the iterative solution should be terminated without waiting for the loop to expire.

In determining when the residual is sufficiently small, it must be kept in mind that we are talking actually about the magnitude of the residual. A residual of -10^{10} is less than 10^{-10}, but it is not small. Thus, it is necessary to consider absolute values in all tests, and absolute values are relatively slow processes from a computational standpoint. A better alternative is to test the squares of the residuals. This eliminates the need to take absolute values, and can provide exactly the same test criterion. An even better test for most purposes is to determine acceptability based on the sum of the squares of the residuals. This quantity can only be small provided *all* of the individual residuals are small, and it can be calculated quite rapidly in the computer.

The second possible termination condition is that the solution does not move. If the solution estimates are of order 10^1 and the calculation is being done to 14 decimal places, then a computed correction of $1 \cdot 10^{-15}$ has no effect when added to the current estimate. It is insignificantly small compared to the current estimates for the variables, and hence, the solution "does not move." To evaluate in the fastest possible manner the possibility that this situation has been reached, it is recommended that the sum of the squares of the components of the vector $\{\Delta x\}$ be tested against an established minimum step. Note that this does not have to be the smallest possible step based on computational concerns. The analyst may simply decide that if the adjustments all drop below a set level that will be taken as an acceptable solution. Again, the analyst must exercise judgment in this matter.

The final possibility is that the residuals are not small and the adjustments are not small either, in which case, another iteration is made. If this situation continues until the loop expires, there will be solution estimate that results. It may be a

Computer-Aided Mathematics for Science and Engineering

reasonable approximation, nearly, but not quite yet, satisfying the other termination criteria, or it may be far removed from a true solution. It is important to be certain why the iteration has ended!

Example

For the purposes of an example, consider the following two equations to be solved for x and y,

$$
\begin{aligned}
x^2 + 3xy - y &= 0 \\
\sin x + y &= 4
\end{aligned}
$$

where it is known that $0 \le x \le 1$ and $0 \le y \le 5$.

Following the process indicated above, let $x_1 = x$ and $x_2 = y$, and then rewrite these equations in the forms $f_1 = 0$ and $f_2 = 0$:

$$
\begin{aligned}
f_1(x, y) &= x^2 + 3xy - y = 0 \\
f_2(x, y) &= \sin x + y - 4 = 0
\end{aligned}
$$

The Jacobian is next developed, as shown above:

$$
\begin{bmatrix} \frac{\partial f_1}{\partial x} & \frac{\partial f_1}{\partial y} \\ \frac{\partial f_2}{\partial x} & \frac{\partial f_2}{\partial y} \end{bmatrix} = \begin{bmatrix} 2x + 3y & 3x - 1 \\ \cos x & 1 \end{bmatrix}
$$

As a first estimate of the solution, assume the middle of each of the intervals given, i.e., assume $x_{(1)} = 0.5$ and $y_{(1)} = 2.5$, and using these assumed values, evaluate the residuals:

$$
\begin{aligned}
f_1\left(x_{(1)}, y_{(1)}\right) &= 1.5 \\
f_2\left(x_{(1)}, y_{(1)}\right) &= -1.0205745
\end{aligned}
$$

This clearly is not the solution. Therefore, an adjustment is calculated as described above, beginning with the evaluation of the Jacobian matrix at the current solution estimate:

$$
\begin{bmatrix} 2x + 3y & 3x - 1 \\ \cos x & 1 \end{bmatrix}\Bigg|_{\substack{x=0.5 \\ y=2.5}} = \begin{bmatrix} 8.5 & .5 \\ .87758256 & 1.0 \end{bmatrix}
$$

The adjustment is then computed,

$$
\begin{aligned}
\begin{Bmatrix} \Delta x \\ \Delta y \end{Bmatrix} &= -\begin{bmatrix} 8.5 & .5 \\ .87758256 & 1.0 \end{bmatrix}^{-1} \begin{Bmatrix} 1.5 \\ -1.0205745 \end{Bmatrix} \\
&= \begin{Bmatrix} -.24937789 \\ 1.2394242 \end{Bmatrix}
\end{aligned}
$$

102

and finally the adjustment is applied to the solution estimate:

$$
\left\{ \begin{array}{c} x_{(2)} \\ y_{(2)} \end{array} \right\} = \left\{ \begin{array}{c} x_{(1)} \\ y_{(1)} \end{array} \right\} + \left\{ \begin{array}{c} -0.24937789 \\ 1.2394242 \end{array} \right\}
$$

$$
= \left\{ \begin{array}{c} 0.25062211 \\ 3.7394242 \end{array} \right\}
$$

Continued iteration shows the actual solution as, approximately,

$$
\left\{ \begin{array}{c} x \\ y \end{array} \right\} = \left\{ \begin{array}{c} 0.32383872 \\ 3.6817919 \end{array} \right\}
$$

for which the residuals are, approximately,

$$
\left\{ \begin{array}{c} f_1(x,y) \\ f_2(x,y) \end{array} \right\} = \left\{ \begin{array}{c} -1.5 \times 10^{-14} \\ 0.0 \end{array} \right\}
$$

A single iteration certain did not provide an adequate solution for most purposes, but note that it did significantly move the solution estimates to values much closer to the final estimates.

An example of the code required for a Newton-Raphson solution is shown below. It should be noted that all of the output and the first few lines (option nolet, option base 1, etc.) can be eliminated if this code is to be embedded in the code for a larger problem.

Example Code for the Newton-Raphson Method

```
! An example of the Newton-Raphson solution technique,
! particularly showing the compact code possible using
! matrix notation
option nolet
option base 1
dim xx(2),ff(2),dd(2),jac(2,2),jaci(2,2)

! starting estimates
xx(1)=0.5                      ! initial estimate for x
xx(2)=2.5                      ! initial estimate for y

itmax=25                       ! maximum iterations allowed
flag=0                         ! indicator of termination mode
for iter=1 to itmax            ! Newton-Raphson iteration loop
    call feval
    fsq=ff(1)^2+ff(2)^2
    if fsq<1e-16 then          ! test squared norm
       flag=1                  ! indicates small residual
       exit for
```

```
      end if
      call jacob
      mat dd=jaci*ff
      mat xx=xx-dd                      ! apply the adjustment
      dsq=dd(1)^2+dd(2)^2
      if dsq<1e-18 then                 ! test squared adjustment
         flag=2                         ! indicates small adjustment
         exit for
      end if
next iter

if flag=0 then
   print "Iteration limit reached, solution not converged."
   print "Current estimates:"
   print "      x = ";xx(1)
   print "      y = ";xx(2)
   stop
else if flag=1 then
   msg$="Iteration stopped for small residual"
else if flag=2 then
   msg$="Iteration stopped for small adjustment"
end if
print "Solution converged after ";iter;" iterations"
print "      x = ";xx(1)
print "      y = ";xx(2)
print "    f1 = ";ff(1)
print "    f2 = ";ff(2)
print msg$

sub feval                              ! evaluation of residuals
   x=xx(1)
   y=xx(2)
   ff(1)=x^2+3*x*y-y
   ff(2)=sin(x)+y-4
end sub

sub jacob                              ! evaluation of Jacobian
   jac(1,1)=2*x+3*y
   jac(1,2)=3*x-1
   jac(2,1)=cos(x)
   jac(2,2)=1
   mat jaci=inv(jac)
end sub

end
```

This code produced the final results given above in only four iterations. This very rapid convergence is typical of the Newton-Raphson method in most cases. It should be noted also, that this exact code can be used for a problem of any size simply by changing the dimension statements and rewriting the two subroutines are required for the particular problem.

An Extension to the Newton-Raphson Method

One aspect of the Newton-Raphson method that causes a problem at times is the need to evaluate the Jacobian matrix. For systems that involve many unknowns, the development of expressions for each of the many terms in the Jacobian can become a laborious analytical task, and then these expressions must be correctly programmed for computer evaluation — a lot of work in some cases! There is an alternative available, the finite difference approximation to the Jacobian matrix.

Recall the form of the Jacobian matrix evaluated at the current solution estimated $\{x_{(i)}\}$

$$
\begin{bmatrix}
\frac{\partial f_1}{\partial x_1} & \frac{\partial f_1}{\partial x_2} & \frac{\partial f_1}{\partial x_3} & \cdots \\
\frac{\partial f_2}{\partial x_1} & \frac{\partial f_2}{\partial x_2} & \frac{\partial f_2}{\partial x_3} & \cdots \\
\frac{\partial f_3}{\partial x_1} & \frac{\partial f_3}{\partial x_2} & \frac{\partial f_3}{\partial x_3} & \cdots \\
\vdots & \vdots & \vdots & \ddots
\end{bmatrix}
$$

A finite difference approximation to the jth column is obtained by

$$
\begin{Bmatrix}
\frac{\partial f_1}{\partial x_j} \\
\frac{\partial f_2}{\partial x_j} \\
\frac{\partial f_3}{\partial x_j} \\
\vdots
\end{Bmatrix}
\approx
\frac{1}{\Delta x_j}
\begin{Bmatrix}
f_1(x_{1(i)}, x_{2(i)}, \ldots x_{j(i)} + \Delta x_j, \ldots) - f_1(x_{1(i)}, x_{2(i)}, \ldots x_{j(i)}, \ldots) \\
f_2(x_{1(i)}, x_{2(i)}, \ldots x_{j(i)} + \Delta x_j, \ldots) - f_2(x_{1(i)}, x_{2(i)}, \ldots x_{j(i)}, \ldots) \\
f_3(x_{1(i)}, x_{2(i)}, \ldots x_{j(i)} + \Delta x_j, \ldots) - f_3(x_{1(i)}, x_{2(i)}, \ldots x_{j(i)}, \ldots) \\
\vdots
\end{Bmatrix}
$$

It could be argued that this is not a very sophisticated approximation, and that is certainly true, but if the purpose is in fact the solution of the system of equations, then this usually works well enough provided that the Δx_j is properly chosen. There is no general approach that always works, but some general guidance can be given:

1. The Δx_j must be sufficiently large that a the difference in f-values is meaningful.

2. It sometimes works to take $\Delta x_j = c x_j$, where c is typically $0.005 \leq c \leq 0.05$.

3. The increment suggested just above does not work well if $x_j \approx 0$, so that in such case it is necessary to assign $\Delta x_j = c'$, where c' is a small number with respect to the scale of the problem but still large enough to satisfy the first consideration above.

This approximation is simply the idea that we can locally replace the actual function by a straight line (a plane parallel to all axes except one in a multidimensional space), and then estimate the solution based on the place where that straight line cuts the axis. It gives a broader understanding to see this as an alternative evaluation of the Jacobian matrix rather than as a separate method of solution, although *regula falsi* is often considered as separate and distinct from Newton-Raphson.

An application of this approach occurs in situations where the function forms are nested, that is, the problem looks like

$$f_1(u, v, w) = 0$$
$$f_2(u, v, w) = 0$$
$$f_3(u, v, w) = 0$$

$$u = u(x, y, z)$$
$$v = v(x, y, z)$$
$$w = w(x, y, z)$$

and the required solution consist in finding the appropriate values of x, y, and z. In practice, the interrelations are often quite involved, and it may be difficult to clearly identify the functions u, v, and w. It is assumed, however, that for any chosen value of x, y, and z, values of f_1, f_2, and f_3 can be computed by whatever means are required.

The Jacobian matrix that is needed for the solution of this problem is

$$\begin{bmatrix} \frac{\partial f_1}{\partial x} & \frac{\partial f_1}{\partial y} & \frac{\partial f_1}{\partial z} \\ \frac{\partial f_2}{\partial x} & \frac{\partial f_2}{\partial y} & \frac{\partial f_2}{\partial z} \\ \frac{\partial f_3}{\partial x} & \frac{\partial f_3}{\partial y} & \frac{\partial f_3}{\partial z} \end{bmatrix}$$

and, although there is no conceptual problem with expressing the elements of this matrix, the labor may be prohibitive. It is not very difficult, however, to write the necessary computer code to approximate the Jacobian as described above; therefore, the solution can be implemented in the computer with relatively little human effort. It should be mentioned, however, that this may consume substantial amounts of computer time. In one particular case, the author repeatedly solved a system of the sort with twenty unknowns, and each solution involved about twelve iterations. Each solution took approximately 50 minutes on a 80486, 33 MHz computer.

References

Conte, S.D., *Elementary Numerical Analysis*, McGraw-Hill, 1965.

Doughty, S., *Mechanics of Machines*, John Wiley & Sons, 1988.

Chapter   10

Ordinary Differential Equations

Since the time of Newton and Leibnitz, differential equations have been the language of much of science and engineering. They are at once very powerful and extremely frustrating. Their power comes from their wide range of applications and the ability to describe continuously varying processes. The frustration comes from the fact that after several centuries of work, only a relatively small set of solutions are known and there are few broad rules for obtaining them.

One of the principal objectives for this chapter is to provide a tool to be used to overcome that frustration. Numerical solution methods for ordinary differential equations provide a means to obtain a solution for many situations where no closed form solution is known. A working relationship with these methods can go far toward removing the anxiety many people feel with regard to differential equations.

The Euler Step

The most basic concepts for numerical solution of ordinary differential equations are found in a method known as the Euler step. It should be stated here that this method is of little value in practice, but it helps the user understand the processes for other, much more useful algorithms. Consider a first-order ordinary differential equation with an initial condition:

$$\dot{y} = f(t, y), \qquad y(0) = y_0$$

Imagine that you confidently set out to draw a graph of the solution for this differential equation. The axes are drawn and labeled, t-axis horizontally, y-axis vertical. The initial value is plotted on the vertical axis at the point $(0, y_0)$. Now, what comes next? It seems that there is no more information about the solution curve, but perhaps that is not quite true. The initial value can be applied to the differential equation to obtain the slope of the curve at the initial point, $\dot{y}(0)$:

$$\dot{y}(0) = f(0, y_0)$$

On the solution graph, draw a short, straight line with slope $\dot{y}(0)$ at the point $(0, y_0)$. Taken together, this says that at time equal to zero, the solution value is $y(0)$ and the slope is $\dot{y}(0)$, both known values. Now the available information has truly been exhausted, and the time has come to use reason (imagination?).

If a value is required for $y(t)$ for some small, positive value of t, often denoted as h, it can be obtained by a Taylor expansion as

$$y_1 = y(h) \quad = y(0) + h\dot{y}(0) + \frac{h^2}{2}\ddot{y}(0) + \cdots$$
$$\approx y(0) + h\dot{y}(0)$$

where the approximation is justified provided that the step size, h, is sufficiently small. For the purpose of drawing the graph of the solution, let this value be accepted and draw a straight line segment joining $(0, y_0)$ and (h, y_1), where y_1 is simply a short notation for the estimated value of $y(h)$. Over the interval $(0, h)$, this straight line segment is the approximate solution.

To carry this process a second step, it is first necessary to obtain the slope at the new point by resorting again to the differential equation:

$$\dot{y}(h) = f(h, y_1)$$

and from this a new point at $t = 2h$ can be estimated, again using a Taylor expansion:

$$y_2 = y(2h) \quad = y_1 + h\dot{y}(h) + \frac{h^2}{2}\ddot{y}(h) + \cdots$$
$$\approx y_1 + h\dot{y}(h)$$

Clearly, this process can be repeated time after time, obtaining estimates for successive points on the curve. The general relation, applicable to finding all later points is

$$y_{i+1} = y_i + h\dot{y}(t_i, y_i)$$

where the derivative is written with two arguments to indicate that t_i and its associated solution value, y_i, are used in $f(t, y)$ to evaluate $\dot{y}(t_i)$.

Note that after the (exact) initial value, all subsequent points are only estimates because the calculation is only approximate. It seems evident that the quality of the approximation will depend on the size of h, and that as h is made smaller, the quality of the estimate is improved. The other side of this coin is that as h is made smaller, many more steps are required to reach any particular event. This is a definite limitation, both in terms of time and in terms of accumulating round-off error. *This method is not recommended for actual application.*

There are numerous refinements of the Euler step, some of which are of interest from the standpoint of numerical analysis and studies of error propagation, but most of the simple methods are of little value for the solution of a differential equation as a means to an end, not the end in itself. Of the methods used for serious solution work, there are two major families:

- Runge-Kutta Methods, which use the current point (t_i, y_i) as the only basis for stepping to the next point, and

- Predictor-Corrector Methods, which employ the current point (t_i, y_i) and previous points, such as (t_{i-1}, y_{i-1}), $(t_{i-2}, y_{i-2}) \ldots$

There is much that appeals about the Predictor-Corrector idea. It seems reasonable and prudent to make use of more than one point for stepping to the new point. There is, however, an obvious difficulty with this approach, namely, "How does it get started?" The usual answer is that a Runge-Kutta starter is used to determine the first few points, and then the solution is switched to the Predictor-Corrector for the remainder of the solution. One can also argue that for an initial value problem, the calculation of the first few points can possibly destroy the validity of the entire solution; therefore, any method that is satisfactory for starting is probably adequate for the continuation as well. This is a bit over simplified, but there is a germ of truth in it as well. Predictor-Corrector methods will not be further considered here, and the user is referred to any of the many standard texts in numerical analysis for more information. The focus below will be on Runge-Kutta solutions.

The word *order* is often used when referring to various members of the Runge-Kutta family, and it is important to be clear about its meaning. The relation used to make the step from one time point to the next will often be written as

$$y_{i+1} = y_i + (\ldots) + O\left(h^{n+1}\right)$$

and this is called an nth order method. As before, the quantity h is the step size or Δt. The notation $O\left(\ldots\right)$ denotes the remainder term from a Taylor expansion of the solution, and the terms preceding this term agree with the Taylor expansion exactly. One of the most common Runge-Kutta methods is the fourth-order Runge-Kutta algorithm that agrees exactly with the Taylor expansion through terms involving h^4 and has an error term proportional to h^5.

Runge-Kutta Methods

The names of quite a number of people are associated with the Runge-Kutta algorithms. The principal names are those of two German applied mathematicians, Carl Runge (1856-1927) and M.W. Kutta (1867-1944). Runge worked at the university in Göttingen from 1904 to 1925. Kutta is otherwise known for his contribution to aerodynamics (the Kutta-Joukowski theory of lift). Others have also contributed to the development of this family of solution methods, notably E.J. Nyström (1925) from Finland, and S. Gill (1951). All of these people did their work in a time before high speed digital computation and deserve our great respect for their accomplishments in an era of manual or mechanical computation.

Consider, as before, the basic problem in the form

$$\dot{y} = f\left(t, y\right) \qquad y\left(0\right) = y_0$$

Runge-Kutta methods of a specified order, say n, always involve a step formula of the form

$$y_{i+1} = y_i + c_o k_o + c_1 k_1 + \cdots + c_n k_n$$

which is required to agree with the Taylor series expansion of the solution through the nth order term. The individual k's are calculated from the differential equation according to

$$
\begin{aligned}
k_0 &= hf\left(t_i, y_i\right) \\
k_1 &= hf\left(t_i + \alpha_1 h, y_i + \beta_{10} k_0\right) \\
k_2 &= hf\left(t_i + \alpha_2 h, y_i + \beta_{20} k_0 + \beta_{21} k_1\right) \\
&\ \ \vdots \\
k_n &= hf\left(t_i + \alpha_n h, y_i + \beta_{n0} k_0 + \beta_{n1} k_1 + \cdots + \beta_{nn-1} k_{n-1}\right)
\end{aligned}
$$

where the c's, α's, and β's are chosen to accomplish the required agreement with the Taylor expansion. The choices for c's, α's, and β's are not unique; there are infinitely many possible choices. This leaves room for choosing the coefficients with a view toward optimizing some other aspect of the numerical solution, and this is what has led to a variety of Runge-Kutta algorithms (even if only one order, such as the fourth, is considered). These include the Runge-Kutta-Nyström algorithm (recommended for improved accuracy), the Runge-Kutta-Gill algorithm (recommended for minimizing required storage), etc. The labor required in the detailed derivation of any of these is great, but the user may find a simple case developed in the book by Hildebrand.

In spite of the range of possibilities, there are only a few Runge-Kutta algorithms that are in frequent use. The classic fourth-order Runge-Kutta (without any other names attached) is as follows:

$$
\begin{aligned}
k_1 &= hf\left(t_i, y_i\right) \\
k_2 &= hf\left(t_i + \frac{1}{2}h, y_i + \frac{1}{2}k_1\right) \\
k_3 &= hf\left(t_i + \frac{1}{2}h, y_i + \frac{1}{2}k_2\right) \\
k_4 &= hf\left(t_i + h, y_i + k_3\right)
\end{aligned}
$$

$$
y_{i+1} = y_i + \frac{1}{6}\left(k_1 + 2k_2 + 2k_3 + k_4\right)
$$

This is a true workhorse, an algorithm that, while probably not optimum in any particular respect, is rugged and adaptable enough to give serviceable answers to a great many problems. The discussion below deals with the details of computer implementation, and while the form outlined applies to all such algorithms, it may be useful to have the classic Runge-Kutta in mind for that discussion.

The outline below shows the steps required in a computer program implementing a fourth-order Runge-Kutta. Note that there are two initialization phases included, one for the program itself and the second for the problem. The first of these includes dimension statements, buffer designations for printers and plotters, and other controls necessary for the execution of the program but not actually related to the particular problem. The second initialization phase has to do directly with the problem and the numerical solution, such things as initial values, step size, and computed constants. Following this, the program moves into the main loop in which the calculations are made for the Runge-Kutta step. The program draws on a subroutine

110

for values of $\dot{y}$ with various arguments as required. After evaluating the k's, the final step within the loop is the update equation, the equation by which the new solution estimate is computed. At that same point, the time (or other independent variable) is advanced by h.

Outline for Typical, Fourth Order Runge-Kutta Program

```
1.  Program Initialization
        Array dimensions, printer selection, plotter selection, etc.


2.  Problem Initialization
        Set initial conditions, initial time, step size, etc.


3.  Begin Runge-Kutta Solution Loop
        a. First Evaluation of Derivative
            - Call Derivative Subroutine
            - calculate k1
            - Print, Plot, and/or Save Results,
                Including Derivative Value If Required
            - Test for Termination
                If positive, exit from Runge-Kutta Loop
        b. Second Evaluation of Derivative
            - Call Derivative Subroutine
            - calculate k2
        c. Third Evaluation of Derivative
            - Call Derivative Subroutine
            - calculate k3
        d. Fourth Evaluation of Derivative
            - Call Derivative Subroutine
            - calculate k4
        e. Solution Update
            Advance Solution & Independent Variable
    End Loop
4.  Subroutine for Derivative Evaluation
        . . . . .
    End Sub
```

Within the main loop, just after the first call to the derivative subroutine, the results are recorded. This may be by sending them to the printer, by plotting another solution point on a graph of the solution, or by storing the data in arrays for other use later. At this point, values are available for the solution and the derivative at the latest solution point, even at the first point on the solution. For this reason, this is an opportune place to record the results. It may also be necessary to make some auxiliary calculations at this point, evaluating additional variables not involved in the differential equation but dependent on the differential equation solution values.

A second matter is also considered at this same point, that of termination. When should the solution process be stopped? The answer depends entirely on the problem being solved and what is being sought. For some classes of problems, it is sufficient to stop when the independent variable reaches a particular value. In a ballistics problem, it might be appropriate to stop whenever the trajectory intersects the ground. In an aircraft pursuit problem, the solution continues until (1) the pursuer is within a specified distance of the target, the kill distance, or (2) the separation exceeds another specified amount, the maximum distance at which the target can be tracked. It is necessary to devise an appropriate test to determine when the solution should end, and the appropriate place to make the test is just after the results are recorded, all as shown in the outline.

Systems of First Order Differential Equations

The extension of the classic fourth-order Runge-Kutta algorithm to deal with several simultaneous first order differential equations is quite direct. Consider specifically a system of two first order equations

$$\begin{aligned} \dot{x} &= f(t, x, y) & x(0) &= x_0 \\ \dot{y} &= g(t, x, y) & y(0) &= y_0 \end{aligned}$$

The expanded form for the classic algorithm is shown below:

$$\begin{aligned}
k_1 &= hf(t_i, x_i, y_i) \\
m_1 &= hg(t_i, x_i, y_i) \\
k_2 &= hf\left(t_i + \frac{1}{2}h, x_i + \frac{1}{2}k_1, y_i + \frac{1}{2}m_1\right) \\
m_2 &= hf\left(t_i + \frac{1}{2}h, x_i + \frac{1}{2}k_1, y_i + \frac{1}{2}m_1\right) \\
k_3 &= hf\left(t_i + \frac{1}{2}h, x_i + \frac{1}{2}k_2, y_i + \frac{1}{2}m_2\right) \\
m_3 &= hg\left(t_i + \frac{1}{2}h, x_i + \frac{1}{2}k_2, y_i + \frac{1}{2}m_2\right) \\
k_4 &= hf(t_i + h, x_i + k_3, y_i + m_3) \\
m_4 &= hg(t_i + h, x_i + k_3, y_i + m_3)
\end{aligned}$$

$$\begin{aligned}
x_{i+1} &= x_i + \frac{1}{6}(k_1 + 2k_2 + 2k_3 + k_4) \\
y_{i+1} &= y_i + \frac{1}{6}(k_1 + 2k_2 + 2k_3 + k_4)
\end{aligned}$$

The extension to larger systems is obvious, but it becomes clumsy to write out. Matrix notation, in which all of the dependent variables comprise a column vector, can be useful at this point. This will be demonstrated in a later section.

Improved Accuracy and Variable Step Size

The common perception is that the solution accuracy increases as the step size is decreased. This is correct except when decreasing step size leads to a significant increase in the number of steps required to reach termination. If the number of steps increases substantially, the round-off error made at each step can become a significant factor adversely affecting the solution accuracy. This raises the who question of what step size should be used in the first place.

If the numerical solution is advanced by a step h, the result is a value that may be designated as y_h. Now consider advancing the solution over that same interval in two steps of $h/2$ each, leading to a value after two half steps designated as $y_{h/2}$. If the difference $\left| y_h - y_{h/2} \right|$ is sufficiently small (here the user must decide what is an acceptable difference) then the step size h is small enough. If, on the other hand, the solution is advanced by $2h$, taken in the first case as a single step with the result y_{2h} and again in two steps of h each with the result y_{h+h}, then the step size may be doubled, provided that the difference $\left| y_{2h} - y_{h+h} \right|$ is acceptably small.

This decision to decrease, increase, or maintain the step size may in principle be made at every step, but in practice it is common to use a fixed step size for the entire solution. The usual approach is to generate a solution with an intuitively determined (fixed) step size, and then repeat the solution using half that value, and finally to compare the two results. There are times, however, when this is not a very satisfactory approach. Consider a situation in which the solution is changing very slowly over one part of the solution time interval and rapidly over another part of the interval. A step size that is small enough to be satisfactory during the interval of rapid change will be unnecessarily short during the interval of slow change. Conversely, a step size that is appropriate to the interval of slow change will likely lose the solution completely when the interval of rapid change is encountered. This situation clearly calls for a variable step size algorithm.

One attractive prospect is the fifth-order Runge-Kutta-Feldberg algorithm discussed below. This algorithm requires six derivative evaluations per step, but from those six evaluations both fourth- and fifth-order Runge-Kutta steps can be made. The difference between these two is an indication of the size of the error that would be made if the fourth order step were made. Based on the size of this difference, called the *local error*, a decision can be made as to whether the step size can be increased, must be shortened, or should be simply maintained. The fifth-order solution is the one carried forward, so this offers improved accuracy both in terms of a simply approach to using a variable step size and in terms of a higher order algorithm that agrees with the Taylor expansion through the h^5 term.

In the listing that follows, note that the solution is stored in the two arrays `tsav()` and `ysav()`. It is necessary to store the time values as well as the solution values because the solution points are no longer evenly spaced in time.

Fifth-Order Runge-Kutta-Feldberg Algorithm

With Variable Step Size

```
! The Runge-Kutta-Fehlberg 5th Order Solution for an ODE
! Including Variable Step Size
option base 0
dim tsav(2500),ysav(2500)

! Initial conditions
indx=0
tsav(indx)=0
ysav(indx)=....
h=....                           ! Initial step size
allerr=....                      ! Allowable error

! R-K-F algorithm loop
do
   ! 1st Evaluation of Derivative
   t=tsav(indx)
   y=ysav(indx)
   call deriv(t,y,ydot)
   k1=2*ydot

   ! Print solution, including derivative, here
   print tsav(i),ysav(i),ydot

   ! Test for termination
   if ... then exit do

   ! 2nd Evaluation of Derivative
   t=tsav(indx)+h/4
   y=ysav(indx)+h*k1/4
   call deriv(t,y,ydot)
   k2=ydot

   ! 3rd Evaluation of Derivative
   t=tsav(indx)+3*h/8
   y=ysav(indx)+(3*k1+9*k2)*h/32
   call deriv(t,y,ydot)
   k3=ydot

   ! 4th Evaluation of Derivative
   t=tsav(indx)+12*h/13
   y=ysav(indx)+(1932*k1-7200*k2+7296*k3)*h/2197
```

114

```
call deriv(t,y,ydot)
k4=ydot

! 5th Evaluation of Derivative
t=tsav(indx)+h/2
y=ysav(indx)+(439*k1/216-8*k2+3680*k3/513-845*k4/4104)*h
call deriv(t,y,ydot)
k5=ydot

! 6th Evaluation of Derivative
t=tsav(indx)+h/2
y=ysav(indx)-(8*k1/27-2*k2+3544*k3/2565)*h
call deriv(t,y,ydot)
k6=ydot

! Solution Update & Error Estimate
dy5=(16*k1/135+6656*k3/12825+28561*k4/56430-9*k5/50+2*k6/55)*h
err=(k1/360-128*k3/4275-2197*k4/75240+k5/50+2*k6/55)*h

    if abs(err)<=allerr then          ! Accept the step
        indx=indx+1                   ! Advance the index
        tsav(indx)=tsav(indx-1)+h     ! Advance the time
        ysav(indx)=ysav(indx-1)+dy5   ! Advance the solution
        if abs(err)<allerr/2 then h=2*h ! Increase step size
    else if abs(err)>allerr then
        h=h/2                         ! Decrease step size
    end if
loop

end

sub deriv(t,y,ydot)
    .....
end sub
```

Near the end of the Runge-Kutta loop, in the `Solution Update & Error Estimate` section, note that two items are evaluated: `dy5` and `err`. The first is the increment in the fifth-order solution and the second is the estimated local error. The absolute value of the local error is then tested against an assigned allowable error (`allerr`), and the decision to increase, decrease, or maintain the step size is made on that basis. If the local error is less than or equal to the allowable error, the solution step just evaluated is accepted, and `dy5` is added to `ysav` to produce a new value of `ysav`. Only if the magnitude of the local error exceeds the allowable error is the last step discarded and those calculations made again with a smaller step size.

Application of Runge-Kutta to $\ddot{y}=f(t,y,\dot{y})$

All of the discussion to this point has dealt with first-order differential equations, but in physical applications, second order differential equations are quite common. These frequently come from applications of Newton's Second Law of motion for a mechanical system or from the loop equations governing an LRC circuit. How can these ideas be applied to solve a second-order differential equation?

Consider the second-order differential equation,

$$\ddot{y} = f\left(t, y, \dot{y}\right) \qquad\qquad y\left(0\right) = y_0, \quad \dot{y}\left(0\right) = \dot{y}_0$$

which is typical of this class of problems. Now consider a change of variables defined by

$$y\left(t\right) = u\left(t\right)$$
$$\dot{y}\left(t\right) = v\left(t\right)$$

When this is applied to the differential equation and the initial conditions, the result is

$$\begin{aligned} \dot{u} &= v & u\left(0\right) &= y_0 \\ \dot{v} &= f\left(t, u, v\right) & v\left(0\right) &= \dot{y}_0 \end{aligned}$$

The extension of the Runge-Kutta algorithm to a system of first-order equations was presented in an earlier section, and those results apply here. Nothing further is required.

However, because of the special nature of the first of these two equations, a further simplification is possible, as found in the works of Abramowitz and Stegun, Hildebrand, and Kreyszig. The algorithm reduces to

$$m_1 = \quad hf\left(t_i, y_i, \dot{y}_i\right)$$

$$m_2 = \quad hf\left(t_i + \tfrac{1}{2}h, y_i + \tfrac{1}{2}h\dot{y}_i + \tfrac{1}{8}hm_1, \dot{y}_i + \tfrac{1}{2}m_1\right)$$

$$m_3 = \quad hf\left(t_i + \tfrac{1}{2}h, y_i + \tfrac{1}{2}h\dot{y}_i + \tfrac{1}{8}hm_2, \dot{y}_i + \tfrac{1}{2}m_2\right)$$

$$m_4 = \quad hf\left(t_i + h, y_i + h\dot{y}_i + \tfrac{1}{2}hm_2, \dot{y}_i + m_3\right)$$

$$\dot{y}_{i+1} = \quad \dot{y}_i + \tfrac{1}{6}\left(m_1 + 2m_2 + 2m_3 + m_4\right)$$

$$y_{i+1} = \quad y_i + h\left[\dot{y}_i + \tfrac{1}{6}\left(m_1 + m_2 + m_3\right)\right]$$

This is more easily programmed than the direct application of the Runge-Kutta algorithm to two first-order differential equations. This is illustrated in the listed given below.

Fourth-Order Runge-Kutta

Applied to $\ddot{y}=f(t,y,\dot{y})$ — Scalar Form

```
option base 0
dim ysav(500),vsav(500)

! Set Initial Values
ysav(0)=...                          ! Initial displacement
vsav(0)=...                          ! Initial velocity
h=...                                ! Integration timestep size
ntp=...                              ! Number of time points

! NBS Integrator for 2nd Order ODE
for timestep=0 to ntp

    ! 1st evaluation of derivative
    yy=ysav(i)
    dy=vsav(i)
    tt=t
    call deriv
    m1=h*ddy
    ! Print, plot, and/or save solutions here
    ! Test for termination here
    if ... then exit for

    ! 2nd evaluation of derivative
    yy=ysav(i)+0.5*h*vsav(i)+0.125*m1*h
    dy=vsav(i)+0.5*m1
    tt=t+h/2
    call deriv
    m2=h*ddy

    ! 3rd evaluation of derivative
    yy=ysav(i)+0.5*h*vsav(i)+.125*m2*h
    dy=vsav(i)+0.5*m2
    tt=t+h/2
    call deriv
    m3=h*ddy

    ! 4th evaluation of derivative
    yy=ysav(i)+h*vsav(i)+0.5*m3*h
    dy=vsav(i)+m3
    tt=t+h
    call deriv
```

```
    m4=h*ddy

    ! Update the solution
    ysav(i)=xsav(i)+h*(vsav(i)+(m1+m2+m3)/6)
    vsav(i)=vsav(i)+(m1+2*m2+2*m3+m4)/6
    t=t+h

next timestep

sub deriv
    ! This subroutine evalutates the second derivative,
    !  based on input values for
    !  yy   - current value of solution
    !  dy   - current velocity
    !  tt   - current time
    ddy=...
end sub


end
```

Yet another difficulty arises when the need arises to for numerical solution of a system of second-order differential equations. This happens frequently in multidegree of freedom dynamics problems, multiloop circuit equations, etc., wherever there are several dependent variables described by second-order differential equations. To be specific, consider the situation described by

$$\ddot{y}_1 = f_1\left(t, y_1, y_2, \dot{y}_1, \dot{y}_2\right) \qquad y_1\left(0\right) = y_{10} \qquad \dot{y}_1\left(0\right) = v_{10}$$
$$\ddot{y}_2 = f_2\left(t, y_1, y_2, \dot{y}_1, \dot{y}_2\right) \qquad y_2\left(0\right) = y_{20} \qquad \dot{y}_2\left(0\right) = v_{20}$$

This system may be written in matrix form as follows:

$$\{\ddot{y}\} = \left\{ \begin{array}{c} \ddot{y}_1 \\ \ddot{y}_2 \end{array} \right\} = \left\{ \begin{array}{c} f_1\left(t, \{y\}, \{\dot{y}\}\right) \\ f_2\left(t, \{y\}, \{\dot{y}\}\right) \end{array} \right\} = \{f\left(t, \{y\}, \{\dot{y}\}\right)\}$$

$$\left\{ \begin{array}{c} y_1\left(0\right) \\ y_2\left(0\right) \end{array} \right\} = \left\{ \begin{array}{c} y_{10} \\ y_{20} \end{array} \right\} \qquad \left\{ \begin{array}{c} \dot{y}_1\left(0\right) \\ \dot{y}_2\left(0\right) \end{array} \right\} = \left\{ \begin{array}{c} v_{10} \\ v_{20} \end{array} \right\}$$

The same algorithm presented earlier in this section can be applied in vector (column matrix) form as follows:

118

$$\{m_1\} = \quad h\left\{f\left(t_i, \{y_i\}, \{\dot{y}_i\}\right)\right\}$$

$$\{m_2\} = \quad h\left\{f\left(t_i + \tfrac{1}{2}h, \{y_i\} + \tfrac{1}{2}h\{\dot{y}_i\} + \tfrac{1}{8}h\{m_1\}, \{\dot{y}_i\} + \tfrac{1}{2}\{m_1\}\right)\right\}$$

$$\{m_3\} = \quad h\left\{f\left(t_i + \tfrac{1}{2}h, \{y_i\} + \tfrac{1}{2}h\{\dot{y}_i\} + \tfrac{1}{8}h\{m_2\}, \{\dot{y}_i\} + \tfrac{1}{2}\{m_2\}\right)\right\}$$

$$\{m_4\} = \quad h\left\{f\left(t_i + h, \{y_i\} + h\{\dot{y}_i\} + \tfrac{1}{2}h\{m_2\}, \{\dot{y}_i\} + \{m_3\}\right)\right\}$$

$$\{\dot{y}_{i+1}\} = \quad \{\dot{y}_i\} + \tfrac{1}{6}\left(\{m_1\} + 2\{m_2\} + 2\{m_3\} + \{m_4\}\right)$$

$$\{y_{i+1}\} = \quad \{y_i\} + h\left[\{\dot{y}_i\} + \tfrac{1}{6}(\{m_1\} + \{m_2\} + \{m_3\})\right]$$

In this form, the algorithm may be applied to any number of simultaneous second-order differential equations simply by appropriately dimensioning the arrays. As might be expected, the program execution time goes up rapidly with the number of equations solved.

A listing of a program implementing these ideas follows. Note that the availability of matrix operations is assumed, but even with that, the code is somewhat lengthy because the matrix operations allow only one simple arithmetic operation per line. This particular code has been employed quite effectively in the integration of multiple degree of freedom vibration problems.

Fourth-Order Runge-Kutta
Applied to $\{\ddot{y}\} = \{\, f(\,t, \{y\}, \{\dot{y}\}\,)\,\}$ — Vector Form

```
option base 1
! ndof  = number of degrees of freedom
!       = number of solution variables
! ntp   = number of time points for solution
dim y(ndof),dy(ndof),ddy(ndof),yy(ndof),vv(ndof),aa(ndof)
dim M(ndof,ndof),MI(ndof,ndof),C(ndof,ndof),K(ndof,ndof)
dim m1(ndof),m2(ndof),m3(ndof),m4(ndof),s(ndof)
dim tsav(ntp),ysav(ndof,ntp),vsav(ndof,ntp),asav(ndof,ntp)
dim yyy(ntp),vvv(ntp),aaa(ntp)


! Set ICs
mat yy=zer                          ! Initial displacements
yy(3)=...                           ! Initial non-zero displacement
mat vv=zer                          ! Initial velocities
vv(2)=...                           ! Initial non-zer velocity
tt=0                                ! Initial time
h=...                               ! Time step


! NBS Integrator for 2nd Order Systems - Vector Form
```

119

```
for i=1 to ntp

    ! 1st Derivative Evaluation
    t=tt
    mat y=yy
    mat dy=vv
    call deriv
    mat m1=(h)*ddy
        print tt,yy(1),yy(2),yy(3).....
    ! Save the results
    tsav(i)=tt
    for ii=1 to ndof
        ysav(ii,i)=yy(ii)
        vsav(ii,i)=vv(ii)
        asav(ii,i)=ddy(ii)
    next ii
    ! Test for termination
    if ... then exit for

    ! 2nd Derivative Evaluation
    t=tt+h/2
    mat y=(h/8)*m1
    mat s=(h/2)*vv
    mat y=y+s
    mat y=y+yy
    mat dy=(1/2)*m1
    mat dy=dy+vv
    call deriv
    mat m2=(h)*ddy

    ! 3rd Derivative Evaluation
    t=tt+h/2
    mat y=(h/8)*m2
    mat s=(h/2)*vv
    mat y=y+s
    mat y=y+yy
    mat dy=(1/2)*m2
    mat dy=dy+vv
    call deriv
    mat m3=(h)*ddy

    ! 4th Derivative Evaluation
    t=tt+h
    mat y=(h/2)*m3
    mat s=(h)*vv
```

```
      mat y=y+s
      mat y=y+yy
      mat dy=vv+m3
      call deriv
      mat m4=(h)*ddy

      ! Update the solutions
      mat s=m1+m2
      mat s=s+m3
      mat s=(1/6)*s
      mat s=s+vv
      mat s=(h)*s
      mat yy=yy+s
      mat s=m2+m3
      mat s=(2)*s
      mat s=m1+s
      mat s=s+m4
      mat s=(1/6)*s
      mat vv=vv+s
      tt=tt+h

next i

sub deriv
      ! input data:     y    = vector of displacements
      !                 dy   = vector of velocities
      !                 t    = time for evaluation
      ! Returns:        ddy  = vector of accelerations
      mat ddy=....
end sub
```

References

Abramowitz, M., and Stegun, I.E., eds., *Handbook of Mathematical Functions*, U.S. Government Printing Office, 10th printing, Dec. 1972, p. 897, item 25.5.20. (Caution: Early printings of this book have an error at this point.)

Hildebrand, F.B., *Introduction to Numerical Analysis*, McGraw-Hill, 1956, pp. 233–238.

Kreyszig, E., *Advanced Engineering Mathematics*, 3rd ed., John Wiley & Sons, 1972, pp. 668–672.

Chapter 11

Fourier Series

Fourier series have been one of the major tools for the mathematical analysis of periodic phenomena for many years. For a long time, it was essentially a formal solution process and only rarely carried through to numerical evaluation. One clear example of this is found in the work of Timoshenko where indicated Fourier sums abound, but only limited numerical results are obtained. During the middle 1960s a major new result was presented by Cooley and Tukey, leading to what is today commonly called the Fast Fourier Transform or FFT. This development has renewed interest in Fourier series as a practical means for obtaining useful, numerical results.

What is a Fourier Series?

Consider a periodic function $f(x)$ having period L. Under rather broad conditions, this function can be represented by a trigonometric series of the form

$$f(x) = a_0 + \sum_{n=1}^{\infty} \left[a_n \cos(\frac{2\pi nx}{L}) + b_n \sin(\frac{2\pi nx}{L}) \right]$$

This sort of series is called a *Fourier series*, and it is widely applicable in the study of vibrations, elasticity, electric and magnetic fields, heat transfer, and other areas. the constants a_n and b_n are called the *Fourier coefficients* for $f(x)$. From a purely theoretical view, the fact that the summation runs up to infinity is of no particular difficulty; from an applications perspective this is definitely a problem. This is usually dealt with by the simple expedient of truncating the series, whereby the upper limit is assigned a finite value, say N_2. In some cases, it may be sufficient to take $N_2 = 5$ while in other cases $N_2 = 50$ or more may be required, all depending on the convergence characteristics of the particular series. For all further work here, the finite form is assumed,

$$f(x) = a_0 + \sum_{n=1}^{N_2} [a_n \cos(\frac{2\pi nx}{L}) + b_n \sin(\frac{2\pi nx}{L})]$$

Classical Evaluation of the Coefficients

One of the obvious problems associated with the use of Fourier series is the matter of how the coefficients are to be evaluated. If the expression for $f(x)$ is simply integrated over one period, the integral of each sine and cosine term is zero. This leaves a result that is easily solved for the constant term of the series, a_0:

$$a_0 = \frac{1}{L} \int_0^L f(x) \, dx$$

Seen in this manner, it is evident that the constant term is simply the average value of $f(x)$ taken over one period.

Next consider multiplication of the series expression for $f(x)$ by $\cos(\frac{2\pi mx}{L})$, followed by integration over one period,

$$\int_0^L f(x) \cos(\tfrac{2\pi nx}{L}) \, dx \;\; = \sum_{n=1}^{N_2} \int_0^L a_n \cos(\tfrac{2\pi nx}{L}) \cos(\tfrac{2\pi mx}{L}) \, dx$$

$$+ \sum_{n=1}^{N_2} \int_0^L b_n \sin(\tfrac{2\pi nx}{L}) \cos(\tfrac{2\pi mx}{L}) \, dx$$

Because of the orthogonality properties of the sine and cosine functions, all of the sine-cosine product integrals vanish, as do all of the cosine-cosine product integrals except for one. The surviving integral on the right side of the equation is the one for which $m = n$. The resulting equations can be solved for a_n, with the result

$$a_n = \frac{2}{L} \int_0^L f(x) \cos(\frac{2\pi nx}{L}) \, dx$$

where $n = 1, 2, 3, \ldots N_2$. A similar process of multiplication by $\sin(\frac{2\pi mx}{L})$ and integration over a period leads to an expression for the coefficient b_n,

$$b_n = \frac{2}{L} \int_0^L f(x) \sin(\frac{2\pi nx}{L}) \, dx$$

The classical application of Fourier series requires the closed form evaluation of these integrals in order to determine the coefficients for any function to be represented by a series. Because of the considerable effort required, the series coefficients for many common wave forms have been tabulated in a number of references such as the *CRC Standard Mathematical Tables*.

Numerical Evaluation - Evenly Spaced Data

There are two major challenges associated with the application of Fourier series. The first is the matter discussed above, the determination of the series coefficients for a given function to be represented. The second is the problem of evaluating a function for which the series coefficients are known. All though this latter appears to be obvious (simply add up the terms, or so it would seem), in application this proves to be extremely laborious to the point of bogging down large digital computers. Both of these problems are greatly reduced with the aid of a recursive calculation, here termed the "sequence calculation."

Before getting into the sequence calculation, another important method should be also mentioned. this is the method known as the *Fast Fourier Transform*, or FFT. This method has won great popularity and serves as the basis for many important measuring instruments, known as *FFT Analyzers*. The "Fourier Transform" referred to in the term Fast Fourier Transform is simply the set of Fourier series coefficients and might be appropriately included in the discussion here. The typical FFT algorithm requires that the function to be transformed (the function whose Fourier series coefficients are to be determined) be represented by a set of 2^N *evenly spaced* data points. This means that the FFT is restricted to data sets of, for example, $...32, 64, 128, 512, 1024, ...$ data points. If the function is described by say 100 data points, the FFT is not directly applicable. The requirement that the data be evenly spaced is also a limitation at times. For these reasons and to limit the length of this discussion, the FFT is not discussed further here. See the References for other sources on the FFT.

Original Data Set

As mentioned in the discussion regarding the FFT, the beginning of numerical evaluation for the Fourier series coefficients is a set of discrete data pairs describing the function. Consider the function $f(x)$ defined by the set of $N_1 + 1$ evenly spaced data points, (numbered $0, 1, 2, ...N_1$): (x_0, f_0), (x_1, f_1), $...(x_{N_1}, f_{N_1})$, representing one period of the function. The locations x_0 and x_{N_1} represent the same point in the periodic cycle; they are exactly one period (L) apart. If the function $f(x)$ is continuous at x_0, then f_0 and f_{N_1} will have the same value. If the function is discontinuous at x_0, then f_0 is the limit from the right at x_0, while f_{N_1} is the limit from the left at x_{N_1}.

Determination of the Coefficients

It would appear that a numerical evaluation of the series coefficients could be made by simply applying one of the standard numerical integration algorithms (such as the rectangular rule, trapezoidal rule, Simpson's rule, etc.) to the evaluation of the integrals for a_0, a_n, and b_n above. This approach is indeed workable for the evaluation of a_0, and the trapezoidal rule is employed in the methods presented here. Further reflection on the matter of evaluating a_n and b_n shows that this approach will require a great number of sine and cosine evaluations, a matter relatively time-consuming even in a digital computer.

Two trigonometric identities play an important role in the sequence calculation

$$\cos(n-1)\theta - 2\cos\theta\cos n\theta + \cos(n+1)\theta = 0$$

and

$$\sin(n-1)\theta - 2\cos\theta\sin n\theta + \sin(n+1)\theta = 0$$

If $\sin(n-1)\theta$, $\cos(n-1)\theta$, $\sin n\theta$, $\cos n\theta$ are all known, then these equations offer a means for evaluating $\sin(n+1)\theta$ and $\cos(n+1)\theta$. Recursive evaluations of the necessary sine and cosine functions are much faster than the usual in-line functions used in computer languages. These recursive evaluations will be employed in the determination of a_n and b_n for $n > 0$.

The trapezoidal rule is applicable in the determination of the constant term, a_0:

$$
\begin{aligned}
a_0 &= \tfrac{1}{L} \int_0^L f(x)\, dx \\
&= \tfrac{1}{L} \cdot \tfrac{L}{N_1} \left(\tfrac{f_0+f_1}{2} + \tfrac{f_1+f_2}{2} + \cdots + \tfrac{f_{N_1-1}+f_{N_1}}{2} \right) \\
&= \tfrac{1}{N_1} \left(\sum_{i=0}^{N_1} f_i - \tfrac{1}{2} f_0 - \tfrac{1}{2} f_{N_1} \right)
\end{aligned}
$$

This result is easily implemented in a computer program.

Consider next the determination of the cosine coefficients for the orders $n > 0$, again approached by means of the trapezoidal rule for integration:

$$
\begin{aligned}
a_n &= \tfrac{2}{L} \int_0^L f(x) \cos\left(\tfrac{2\pi n x}{L} \right) dx \\[2ex]
&= \tfrac{2}{N_1} \left[\sum_{i=0}^{N_1} f_i \cos\left(\tfrac{2\pi n i}{N_1} \right) - \tfrac{1}{2} f_0 - \tfrac{1}{2} f_{N_1} \right]
\end{aligned}
$$

The indicated summation calls for the evaluation of a large number of cosines. This is a point where the recursive evaluation is required. Define a sequence F_i such that

$$
F_i = 2\cos\left(\frac{2\pi n}{N_1} \right) F_{i+1} - F_{i+2} + f_i
$$

where $F_{N_1+1} = F_{N_1+2} = 0$. Note that the original data enters as the last term in the calculation of each F_i. The order of evaluation is $F_{N_1}, F_{N_1-1}, \ldots F_0$, where the two starting values, F_{N_1+1} and F_{N_1+2}, were assigned zero values as defined. The sequence definition is solved for the data item, f_i,

$$
f_i = F_i - 2\cos\left(\frac{2\pi n}{N_1} \right) F_{i+1} + F_{i+2}
$$

and the result applied in the summation for a_n:

$$
\begin{aligned}
\sum_{i=0}^{N_1} f_i \cos\left(\frac{2\pi n i}{N_1} \right) =\ & \left[F_{N_1} - 2\cos\left(\frac{2\pi n}{N_1} \right) F_{N_1+1} + F_{N_1+2} \right] \cos(2\pi n) \\
& + \left[F_{N_1-1} - 2\cos\left(\frac{2\pi n}{N_1} \right) F_{N_1} + F_{N_1+1} \right] \cos\left[\frac{2\pi n(N_1-1)}{N_1} \right] \\
& + \cdots + \\
& + \left[F_1 - 2\cos\left(\frac{2\pi n}{N_1} \right) F_2 + F_3 \right] \cos\left(\frac{2\pi n}{N_1} \right) \\
& + \left[F_0 - 2\cos\left(\frac{2\pi n}{N_1} \right) F_1 + F_2 \right] \quad (1)
\end{aligned}
$$

If this is rearranged, with the F's taken as factors out of each group, the summation becomes:

$$\sum_{i=0}^{N_1} f_i \cos\left(\frac{2\pi ni}{N_1}\right) = F_{N_1}\left\{\cos\left(2\pi n\right) - 2\cos\left(\frac{2\pi n}{N_1}\right)\cos\left[\frac{2\pi n(N_1-1)}{N_1}\right] + \cos\left[\frac{2\pi n(N_1-2)}{N_1}\right]\right\}$$

$$+ \cdots +$$

$$+ F_2\left[\cos\left(\frac{4\pi n}{N_1}\right) - 2\cos\left(\frac{2\pi n}{N_1}\right)\cos\left(\frac{2\pi n}{N_1}\right) + 1\right]$$

$$+ F_1\left[\cos\left(\frac{2\pi n}{N_1}\right) - 2\cos\left(\frac{2\pi n}{N_1}\right)\right] + F_0 \cdot (1)$$

The cosine identity reduces each of the terms on the right side to zero, except for the last line. The final result is

$$\sum_{i=0}^{N_1} f_i \cos\left(\frac{2\pi ni}{N_1}\right) = -F_1 \cos\left(\frac{2\pi n}{N_1}\right) + F_0$$

In place of evaluating the sum on the left side with numerous cosine evaluations required, all that is necessary is the recursive evaluation of the sequence for F and a single cosine evaluation! The final expression for a_n is:

$$a_n = \frac{2}{N_1}\left[F_0 - F_1 \cos\left(\frac{2\pi n}{N_1}\right) - \frac{1}{2}f_0 - \frac{1}{2}f_{N_1}\right]$$

The evaluation of the sine coefficients begins with the integral definition,

$$b_n = \frac{2}{L}\int_0^L f(x)\sin(\frac{2\pi nx}{L})\,dx$$

and proceeds through a similar development to the result

$$b_n = \frac{2}{N_1}F_1 \sin\left(\frac{2\pi n}{N_1}\right)$$

The F_1 appearing in this expression is a member of the same sequence defined previously in connection with the cosine coefficients. Thus, for a particular n, evaluation of both a_n and b_n is accomplished with the evaluation of one sine, one cosine, and the sequence for F_i. This can be programmed in a very compact fashion and requires only minimal storage and execution time.

Summing the Series for $f(x)$

The other major task involved in applying Fourier series is that of evaluating the series expression to determine a value for $f(x)$,

$$f(x) = a_0 + \sum_{n=1}^{N_2}[a_n \cos(\frac{2\pi nx}{L}) + b_n \sin(\frac{2\pi nx}{L})]$$

While it appears that this should be a straight forward summation process, there is again the problem of many sines and cosines to be evaluated. The sequence calculation is applicable to this task as well.

The coefficients are now presumed to be known for the series to be summed, and an argument value, a value of x, is specified. Two new sequences are required, defined as

$$A_n = 2\cos\left(\tfrac{2\pi x}{L}\right) A_{n+1} - A_{n+2} + a_n$$

$$B_n = 2\cos\left(\tfrac{2\pi x}{L}\right) B_{n+1} - B_{n+2} + b_n$$

where $A_{N_2+1} = A_{N_2+2} = 0$ and $B_{N_2+1} = B_{N_2+2} = 0$. These sequences are identical in form to that used in the determination of the Fourier coefficients, but note that the final term in each of these sequences is one of the Fourier coefficients. As was done previously, these sequences are evaluated in descending index order, beginning with the evaluation of A_{N_2} and B_{N_2} and working down to A_0 and B_0. The sequence definitions are solved for the Fourier series coefficients, a_n and b_n, and those expressions substituted into the Fourier series. As before, most of the terms will then sum to zero in groups, with the following terms remaining:

$$f(x) = A_0 - A_1 \cos\left(\frac{2\pi x}{L}\right) + B_1 \sin\left(\frac{2\pi x}{L}\right)$$

Thus, evaluation of the series at one argument requires the determination of one sine, one cosine, and the sequences A_i and B_i. If a new argument is considered, all of these must be re-evaluated at that new argument.

Computer Implementation of the Sequence Calculation Approach

The two subroutines listed in this section implement the discussion given above using the sequence calculation to determine Fourier series coefficients and to sum the series. The first subroutine, FSCoef, is applicable only for evenly spaced data values. The second subroutine, SerEval, is applicable to the evaluation of any Fourier series for which the coefficients are known.

Computer Subroutines Based on the Sequence Calculation for
Determining Fourier Series Coefficients for Evenly Spaced Data and
Evaluation of Fourier Series (Summing the Series)

```
sub FSCoef(N1,N2,f(),a(),b(),capF())
    ! a Subroutine for Fourier Series Coefficient Determination
    ! Array dimensions begin at 0
    !       N1 = Max Data Index, Index Runs 0, 1, 2, ... N1
    !       N2 = Max Order Value, n-max
    !     f(i) = Tabulated Function Values - Original Data set
    ! capF(i) = Sequence Values
    !     a(n) = Resulting Vector of Cosine Coef
    !     b(n) = Resulting Vector of Sine Coef

    ! Evaluation of a(0)
    sum=0
    for i=0 to N1
```

```
        sum=sum+f(i)
    next i
    a(0)=(sum-0.5*f(0)-0.5*f(N1))/N1    ! Accounts for discontinuity at ends
    b(0)=0

    ! Evaluation of a(n) & b(n), n>0
    for n=1 to N2
        mat capF=zer
        C=cos(2*pi*n/N1)
        S=sin(2*pi*n/N1)
        for i=N1 to 0 step -1
            capF(i)=2*C*capF(i+1)-capF(i+2)+f(i)
        next i
        a(n)=2*(capF(0)-capF(1)*C-0.5*f(0)-0.5*f(N1))/N1
        b(n)=2*capF(1)*S/N1
    next n
end sub


sub Sereval(N2,x,L,a(),b(),capA(),capB(),fx)
    ! Summation of Fourier Series for Known Coefficients
    ! Array dimensions begin at 0
    !       x = Evaluation Position
    !       L = Length of Period
    !      N2 = Max Order Number
    ! capA() = Sequence values for capA()
    ! capB() = Sequence values for capB()
    !      fx = Computed Function Value - the Result

    mat capA=zer
    mat capB=zer
    C=cos(2*pi*x/L)
    S=sin(2*pi*x/L)
    for n=N2 to 0 step -1
        capA(n)=2*C*capA(n+1)-capA(n+2)+a(n)
        capB(n)=2*C*capB(n+1)-capB(n+2)+b(n)
    next n
    fx=capA(0)-capA(1)*C+capB(1)*S
end sub
```

Numerical Evaluation - Unevenly Spaced Data

There are times when evenly spaced data is not practical. If a function is changing very rapidly over one part of the period and slowly over the rest, it may be necessary to use many closely spaced data points to adequately represent the function during the interval of rapid change and more widely spaced data points were the function is changing slowly.

There is another situation where unevenly spaced data is unavoidable. Suppose that two phenomena are recorded, say $f(t)$ and $g(t)$. Suppose that these are recorded at evenly spaced values of time, say Δt, with the result being two tabular functions, f_0, f_1, ... f_{N_1} and g_0, g_1, ... g_{N_1}. There are certainly cases where it is useful to eliminate the time dependence between the two tabulations and consider these two lists to define a function $f(g)$. Such a tabular relationship will be at unevenly spaced intervals (the g-values). To obtain a Fourier series for this relation will require a means to deal with unevenly spaced data.

As before, consider the function $f(x)$ defined by the set of $N_1 + 1$ unevenly spaced data points, (numbered $0, 1, 2, ...N$): (x_0, f_0), (x_1, f_1), ... (x_N, f_N), representing one period of the function. The locations x_0 and x_N represent the same point in the periodic cycle; they are exactly one period (L) apart. If the function $f(x)$ is continuous at x_0, then f_0 and f_N will have the same value. If the function is discontinuous at x_0, then f_0 is the limit from the right at x_0, while f_N is the limit from the left at x_N.

The process to be presented here consists of three parts: (1) curve fitting the data in groups of three points each (with the possibility of a single pair at the end), (2) closed form evaluation of the integrals defining the Fourier coefficients for each group of three data points, and (3) a recursion algorithm for the evaluation of the integer order sines and cosines required. A listing is given later for a program implementing this approach.

Curve Fit

Consider three consecutive data pairs, (x_i, f_i), (x_{i+1}, f_{i+1}) and (x_{i+2}, f_{i+2}), for which a second-degree curve fit is required of the form

$$y = a + bx + cx^2$$

An exact fit to these three points is made using the coefficients

$$c = \frac{1}{x_{i+2}-x_{i+1}} \left(\frac{f_{i+2}-f_i}{x_{i+2}-x_i} - \frac{f_{i+1}-f_i}{x_{i+1}-x_i} \right)$$

$$b = \frac{f_{i+1}-f_i}{x_{i+1}-x_i} - c(x_{i+1}+x_i)$$

$$a = f_i - x_i(b + cx_i)$$

The validity of this fit can be check by direct substitution.

Closed Form Evaluation of Integrals

The Fourier series coefficients are determined from their integral definitions,

$$a_0 = \frac{1}{L} \int_{x_0}^{x_N} f(x) \, dx = \frac{1}{L} \sum_{i=0,2,4,\ldots}^{N-2} \int_{x_i}^{x_{i+2}} f(x) \, dx$$

$$a_n = \frac{2}{L} \int_{x_0}^{x_N} f(x) \cos\left(\frac{2\pi n x}{L}\right) dx = \frac{1}{L} \sum_{i=0,2,4,\ldots}^{N-2} \int_{x_i}^{x_{i+2}} f(x) \cos\left(\frac{2\pi n x}{L}\right) dx$$

$$b_n = \frac{2}{L} \int_{x_0}^{x_N} f(x) \sin\left(\frac{2\pi n x}{L}\right) dx = \frac{1}{L} \sum_{i=0,2,4,\ldots}^{N-2} \int_{x_i}^{x_{i+2}} f(x) \sin\left(\frac{2\pi n x}{L}\right) dx$$

The objective then, at this point, is the evaluation of the integrals in the right column in closed form.

The first of these is simply the basic step of Simpson's rule integration. Thus

$$\int_{x_i}^{x_{i+2}} f(x) \, dx = \int_{x_i}^{x_{i+2}} \left(a + bx + cx^2\right) dx$$

$$= (x_{i+2} - x_i)\left[a + \frac{b}{2}(x_{i+2} + x_i) + \frac{c}{3}\left(x_{i+2}^2 + x_{i+2}x_i + x_i^2\right)\right]$$

Summing terms of this form will give the constant term, a_0, of the Fourier series by Simpson's rule integration.

For the orders $n > 0$, the integral required for the determination of the cosine coefficient is

$$\int_{x_i}^{x_{i+2}} f(x) \cos\left(\frac{2\pi n x}{L}\right) dx$$

$$= \int_{x_i}^{x_{i+2}} \left(a + bx + cx^2\right) \cos\left(\frac{p_n x}{L}\right) dx$$

$$= a \int_{x_i}^{x_{i+2}} \cos\left(\frac{p_n x}{L}\right) dx$$

$$+ b \int_{x_i}^{x_{i+2}} x \cos\left(\frac{p_n x}{L}\right) dx$$

$$+ c \int_{x_i}^{x_{i+2}} x^2 \cos\left(\frac{p_n x}{L}\right) dx$$

where $p_n = 2\pi n/L$. Each of the integrals above is known, so the evaluation is routine. The result is

$$\int_{x_i}^{x_{i+2}} f(x) \cos\left(\frac{2\pi n x}{L}\right) \, dx = W_{i+2} \sin\left(p_n x_{i+2}\right) + V_{i+2} \cos\left(p_n x_{i+2}\right)$$

$$-W_i \sin\left(p_n x_i\right) - V_i \cos\left(p_n x_i\right)$$

where the functions W and V are

$$W_{i+2} = \frac{1}{p_n}\left(a + bx_{i+2} + c \cdot \frac{(p_n x_{i+2})^2 - 2}{p_n^2}\right) \qquad V_{i+2} = \frac{1}{p_n^2}\left(b + 2cx_{i+2}\right)$$

$$W_i = \frac{1}{p_n}\left(a + bx_i + c \cdot \frac{(p_n x_i)^2 - 2}{p_n^2}\right) \qquad V_i = \frac{1}{p_n^2}\left(b + 2cx_i\right)$$

A similar analysis for the sine coefficients gives the result

$$\int_{x_i}^{x_{i+2}} f(x) \sin\left(\frac{2\pi n x}{L}\right) \, dx = \int_{x_i}^{x_{i+2}} \left(a + bx + cx^2\right) \sin\left(\frac{p_n x}{L}\right) \, dx$$

$$= -W_{i+2} \cos\left(p_n x_{i+2}\right) + V_{i+2} \sin\left(p_n x_{i+2}\right)$$

$$+ W_i \cos\left(p_n x_i\right) - V_i \sin\left(p_n x_i\right)$$

where W and V are as previously defined. This provides a means for determining the contribution to a_n and b_n from each pair of intervals. Although more involved than the results shown earlier for evenly spaced data, this approach is not too difficult to program (as will be seen in the computer code listed below) and has been used effectively.

Recursive Relation for Sines and Cosines

There is clearly a need for the evaluation of quite a few sines and cosines in the procedure described in the previous section. A recursive evaluation that has been found useful for this purpose is based on the trigonometric identities for the cosine and sine of the sum of two angles:

$$\cos\left[\frac{2\pi(n+1)x}{L}\right] = \cos\left(\frac{2\pi n x}{L}\right)\cos\left(\frac{2\pi x}{L}\right) - \sin\left(\frac{2\pi n x}{L}\right)\sin\left(\frac{2\pi x}{L}\right)$$

$$\sin\left[\frac{2\pi(n+1)x}{L}\right] = \sin\left(\frac{2\pi n x}{L}\right)\cos\left(\frac{2\pi x}{L}\right) + \cos\left(\frac{2\pi n x}{L}\right)\sin\left(\frac{2\pi x}{L}\right)$$

Beginning with values for $\cos\left(\frac{2\pi x}{L}\right)$ and $\sin\left(\frac{2\pi x}{L}\right)$, and recalling that $\cos(0) = 1$ and $\sin(0) = 0$, values are readily determined recursively for $\cos(\frac{4\pi x}{L})$, $\cos\left(\frac{6\pi x}{L}\right)$, ... and $\sin(\frac{4\pi x}{L})$, $\sin\left(\frac{6\pi x}{L}\right)$, ...

Subroutine UFSCoef

A subroutine implementing the procedure outlined above is listed below. The user will not find the logic difficult, and all of the steps described above will be

131

easily identified. It should be noted that the subroutine SerEval (listed previously) can be used to sum the series for this case as well; when it comes to summing the series, it matters not whether the original data were evenly or unevenly spaced.

Determination of Fourier Series Coefficients for Unevenly Spaced Data

```
! The array sizes must be at least as large as
! the following three lines indicate ...
! The arrays begin with zero index value
dim f(N1),x(N1)
dim a(N2),b(N2)
dim spnxi(N2),cpnxi(N2),spnxi2(N2),cpnxi2(N2)

sub UFSCoef(N1,N2,x(),f(),spnxi(),cpnxi(),spnxi2(),cpnxi2(),a(),b())
    ! A subroutine for determining Fourier series
    ! coefficients for unevenly spaced data
    !           N1 = Number of data points, numbered 0,1,2,...N1
    !           N2 = Number of coefficients to be determined
    !          x() = Data point abscissa values
    !          f() = Data point function values
    !      spnxi() = Scratch array
    !     spnxi2() = Scratch array
    !      cpnxi() = Scratch array
    !     cpnxi2() = Scratch array
    !          a() = Cosine coeff array
    !          b() = Sine coeff array
    mat spnxi=zer                    ! Assign to zero
    mat cpnxi=(1)                    ! Assign to 1.0
    L=x(N1)-x(0)

    for i=0 to N1-2 step 2
        i2=i
        i1=i+1
        fi=f(i2)
        xi=x(i2)
        fi1=f(i1)
        xi1=x(i1)
        fi2=f(i+2)
        xi2=x(i+2)

        ! Determine coefficients for polynomial approximation
        c4=((fi2-fi)/(xi2-xi)-(fi1-fi)/(xi1-xi))/(xi2-xi1)
        b4=(fi1-fi)/(xi1-xi)-c4*(xi1+xi)
        a4=fi-xi*(b4+c4*xi)
```

```
! Evaluation of sines and cosines for
! interger multiples of the argument
s2pixL=sin(2*pi*x(i+2)/L)
c2pixL=cos(2*pi*x(i+2)/L)
spnxi2(0)=0                      ! Sine of pn*x/L
cpnxi2(0)=1                      ! Cosine of pn*x/L
for n=1 to N2
     spnxi2(n)=spnxi2(n-1)*c2pixL+cpnxi2(n-1)*s2pixL
     cpnxi2(n)=cpnxi2(n-1)*c2pixL-spnxi2(n-1)*s2pixL
next n
a(0)=a(0)+(xi2-xi)*(a4+b4/2*(xi2+xi)+c4/3*(xi2^2+xi2*xi+xi^2))
for n=1 to N2
     pn=2*pi*n/L
     Wi=(a4+b4*xi+c4*((pn*xi)^2-2)/pn^2)/pn
     Wi2=(a4+b4*xi2+c4*((pn*xi2)^2-2)/pn^2)/pn
     Vi=(b4+2*c4*xi)/pn^2
     Vi2=(b4+2*c4*xi2)/pn^2
     a(n)=a(n)+Wi2*spnxi2(n)+Vi2*cpnxi2(n)-Wi*spnxi(n)-Vi*cpnxi(n)
     b(n)=b(n)-Wi2*cpnxi2(n)+Vi2*spnxi2(n)+Wi*cpnxi(n)-Vi*spnxi(n)
next n
mat spnxi=spnxi2
mat cpnxi=cpnxi2
next i
mat a=(2/L)*a
mat b=(2/L)*b
a(0)=0.5*a(0)                    ! 1/L = 0.5*(2/L)
end sub
```

Comparison of Numerical Results

Clearly, there are approximations involved in both of the numerical evaluation schemes presented above, primarily in the numerical integrations. How good are the results? This question does not have a unique answer because the quality of the results is somewhat dependent on the particular problem being addressed. Even so, it is interesting to look at the nature of the results obtained for a typical situation.

For this test case, data has been made up using a Fourier series with known coefficients. The coefficient values are shown in Table 11.1 and the functional form is as follows:

$$f(x) = a_0 + \sum_{n=1}^{6} \left[a_n \cos\left(\frac{2\pi nx}{L}\right) + b_n \sin\left(\frac{2\pi nx}{L}\right) \right]$$

where $L = 1$

Table 11.1 Fourier Series Coefficients

for Function Definition

a_n	b_n

$a_0 = +0.416500$

$a_1 = +0.090900$	$b_1 = -0.312200$
$a_2 = -0.034550$	$b_2 = +0.076200$
$a_3 = -0.006840$	$b_3 = -0.011665$
$a_4 = +0.003690$	$b_4 = -0.004030$
$a_5 = +0.009000$	$b_5 = +0.000000$
$a_6 = +0.009000$	$b_6 = -0.004170$

Table 11.2   Comparison of Cosine Coefficients

Index	a'_n	$a'_n - a_n$	a''_n	$a''_n - a_n$
0	$+0.416500$	$-5.6 \cdot 10^{-17}$	$+0.416500$	$+2.8 \cdot 10^{-16}$
1	$+0.090900$	$-3.5 \cdot 10^{-15}$	$+0.090900$	$-3.8 \cdot 10^{-7}$
2	-0.034550	$+3.0 \cdot 10^{-16}$	-0.034548	$+2.3 \cdot 10^{-6}$
3	-0.006840	$-4.6 \cdot 10^{-16}$	-0.006838	$+2.3 \cdot 10^{-6}$
4	$+0.003690$	$-1.3 \cdot 10^{-17}$	$+0.003686$	$-3.8 \cdot 10^{-6}$
5	$+0.009000$	$-1.6 \cdot 10^{-16}$	$+0.008978$	$-2.2 \cdot 10^{-5}$
6	$+0.009000$	$-2.6 \cdot 10^{-16}$	$+0.008954$	$+4.6 \cdot 10^{-5}$
7	-0.0	$-1.7 \cdot 10^{-16}$	-0.0	$-8.4 \cdot 10^{-17}$
8	-0.0	$-1.4 \cdot 10^{-16}$	$+0.0$	$+4.8 \cdot 10^{-18}$
9	-0.0	$-1.6 \cdot 10^{-16}$	-0.0	$-3.3 \cdot 10^{-16}$
10	-0.0	$-2.1 \cdot 10^{-16}$	-0.0	$-1.1 \cdot 10^{-16}$
11	-0.0	$-4.8 \cdot 10^{-17}$	-0.0	$-8.1 \cdot 10^{-17}$
12	-0.0	$-1.9 \cdot 10^{-16}$	-0.0	$-1.5 \cdot 10^{-16}$
13	-0.0	$-7.0 \cdot 10^{-17}$	-0.0	$-9.3 \cdot 10^{-17}$
14	-0.0	$-1.0 \cdot 10^{-16}$	-0.0	$-3.5 \cdot 10^{-16}$
15	-0.0	$-1.0 \cdot 10^{-16}$	-0.0	$-2.8 \cdot 10^{-16}$

The definition is evaluated at 51, evenly spaced points, (including both end points) on the interval $[0, 1]$ for the comparison. This provides a function for which the Fourier series coefficients are known (by virtue of the defining relation). The point of the test is then to see how well these coefficients are recovered from the data points. Note also that this same data set can be used for both the evenly spaced and unevenly spaced solution techniques. The reason for this is that the unevenly spaced algorithm actually applies in both situations. The true Fourier series coefficients (used in the function definition) are denoted by a_n and b_n in the comparisons shown in Tables 11.2 and 11.3. In these tables, the results from subroutine FSCoef, which is limited to evenly spaced data only, are denoted by a'_n and b'_n. Results obtained from subroutine UFSCoef are denoted by a''_n and b''_n.

Table 11.3   Comparison of Sine Coefficients

Index	b'_n	$b'_n - b_n$	b''_n	$b''_n - b_n$
1	-0.312200	$-1.9 \cdot 10^{-15}$	-0.312199	$+1.3 \cdot 10^{-6}$
2	$+0.076200$	$-5.0 \cdot 10^{-16}$	$+0.076195$	$-5.0 \cdot 10^{-6}$
3	-0.011665	$-1.5 \cdot 10^{-16}$	-0.011661	$+3.9 \cdot 10^{-6}$
4	-0.004030	$-1.0 \cdot 10^{-17}$	-0.004026	$+4.2 \cdot 10^{-6}$
5	-0.000000	$-6.5 \cdot 10^{-17}$	$+0.000000$	$+2.3 \cdot 10^{-16}$
6	-0.004170	$-7.7 \cdot 10^{-17}$	-0.004149	$+2.1 \cdot 10^{-5}$
7	-0.0	$-1.3 \cdot 10^{-16}$	-0.0	$-1.4 \cdot 10^{-16}$
8	-0.0	$-4.5 \cdot 10^{-17}$	-0.0	$-2.4 \cdot 10^{-16}$
9	-0.0	$-1.1 \cdot 10^{-16}$	-0.0	$-5.4 \cdot 10^{-18}$
10	-0.0	$-9.7 \cdot 10^{-17}$	$+0.0$	$+9.6 \cdot 10^{-17}$
11	-0.0	$-5.9 \cdot 10^{-17}$	-0.0	$-1.3 \cdot 10^{-16}$
12	-0.0	$-1.9 \cdot 10^{-16}$	-0.0	$-3.7 \cdot 10^{-17}$
13	-0.0	$-8.0 \cdot 10^{-17}$	$+0.0$	$+1.0 \cdot 10^{-16}$
14	-0.0	$-1.0 \cdot 10^{-16}$	-0.0	$-1.1 \cdot 10^{-16}$
15	-0.0	$-1.5 \cdot 10^{-16}$	$+0.0$	$+1.2 \cdot 10^{-16}$

Application to the Solution of Differential Equations

Fourier series have been widely used in the solution of both ordinary and partial differential equations. The discussion here will be limited to ordinary differential equations, and further limited to the solution of first order, ordinary differential equations with periodic excitation. Consider such an equation

$$\tau \dot{y} + y = f(t) = a_0 + \sum_{n=1}^{N_2} [a_n \cos (n\Omega t) + b_n \sin(n\Omega t)]$$

where

τ is the time constant, and

Ω is the fundamental excitation frequency.

The fundamental frequency is related to the period, T, by $\Omega T = 2\pi$, so that $n\Omega t$ can also be written as $2\pi n t/T$, similar to the expression $2\pi n x/L$ used for spatial variation. Obviously, the same techniques can be used to generate the coefficients and to sum the series without regard for the choice of the independent variable (x or t). One application of Fourier series is evident in the statement of the differential equation; the excitation has been expressed as a Fourier series. A second application will appear shortly below.

The particular solution for the differential equation may be considered as the sum of the particular solutions for each of the terms comprising the excitation,

$$y_p(t) = f_0(t) + f_1(t) + f_2(t) + \cdots$$

where $f_0, f_1, f_2, \ldots$ are each particular solutions associated with one of the excitation terms. For example, consider the constant term of the excitation, a_0, for which the particular solution is a constant, $f_0 = c_0$. Thus,

$$\tau \cdot (0) + c_0 = a_0$$

from which we may conclude that $c_0 = a_0$.

It is convenient to consider the remaining excitation terms in pairs, and for the nth pair the trial form for the solution reads,

$$y_n(t) = c_n \cos(n\Omega t) + d_n \sin(n\Omega t)$$

where c_n and d_n are to be determined. When the trial form is differentiated and substituted into the differential equation, the result reads

$$n\Omega \tau \left[-c_n \sin\left(n\Omega t\right) + d_n \cos\left(n\Omega t\right) \right]$$

$$+ \left[c_n \cos(n\Omega t) + d_n \sin(n\Omega t) \right]$$

$$= a_n \cos(a_n \cos\left(n\Omega t\right) + b_n \sin(n\Omega t)$$

The sine and cosine terms must balance separately leading to the two equations:

$$c_n + n\Omega \tau d_n = a_n$$

$$-n\Omega \tau c_n + d_n = b_n$$

This system of equations is solved for c_n and d_n:

$$c_n = (a_n - n\Omega \tau b_n) / \left[1 + (n\Omega \tau)^2 \right]$$

$$d_n = (n\Omega \tau a_n + b_n) / \left[1 + (n\Omega \tau)^2 \right]$$

If the above process is carried out for each pair of excitation terms, the complete solution can be written as

$$y_p(t) = c_0 + \sum_{n=1}^{N_2} \left[c_n \cos(n\Omega t) + d_n \sin(n\Omega t) \right]$$

which is clearly a Fourier series for the particular solution. Evaluation of this series and determination of the series coefficients for the excitation series both entail the same basic processes described in the previous sections.

A similar process applies for the solution of second-order ordinary differential equations as well. The development of this solution, in a manner parallel to that given here is a very instructive exercise.

Fourier Analysis Difficulties and Errors

The assumption of periodicity is fundamental to all work with Fourier series. When dealing with Fourier series numerically, where functions are represented in tabular form, it is implicitly assumed that the tabulated data represents exactly an integer multiple of the actual fundamental period of the function. Two major error types predominate in dealing numerically with Fourier series: *truncation errors* and *aliasing errors*. These two error types are commonly associated with the FFT, but in fact, they are associated with all discrete (digital) transforms.

Truncation Error

The time interval represented by the data (the data collection interval in the case of experimental data) is referred to as the *window length*. If the window length does not correspond exactly to an integer multiple of the actual fundamental periods of the data, then the data cannot possibly represent an integer number of periods of the original data. Even so, *it does represent exactly one period of some function, the function that will be transformed.* This function looks like the original data in many respects, except near the ends of the window. This replacement function often has discontinuities at the ends of the window. The results, in terms of the series coefficients determined, are

- Component amplitudes are not correct, and

- Many small components ("grass") appear in the spectrum, components that were not present at all in the original function.

Aliasing Error

In order for a data sample to adequately represent a (single frequency) sinusoid, the sampling frequency must be more than twice the frequency of the sinusoid. This is the Nyquist sampling criterion. Note that it states an inequality rather than an equality. Failure to sample at a high enough frequency results in "aliasing," a phenomenon in which components of the spectrum are "folded down" from higher frequencies and erroneously added into the lower frequency spectrum.

The aliasing problem has led to the introduction of "anti-aliasing" filters for conditioning experimental data. These are low-pass filters designed to remove frequencies above the sampling capability of the measurement system. (This creates what is called a "band-limited" function, a function that can be represented exactly by a finite Fourier series.) This must be done with the analog data. Once the data is in digital form, aliasing has already occurred if the data was not properly filtered and it cannot be removed.

References

Acton, F.S., *Numerical Methods That Work*, Harper & Row, 1970.

Brigham, O.E., *The Fast Fourier Transform*, Prentice-Hall, 1974.

Carslaw, H.S., *Introduction to the Theory of Fourier's Series and Integrals*, 3rd. rev. ed., Dover, 1930.

Churchill, R.V., *Fourier Series and Boundary Value Problems*, McGraw-Hill, 1963.

CRC Standard Mathematical Tables, Chemical Rubber Publishing Co. (revised annually).

Cooley, J.W., and Tukey, J.W., "An Algorithm for the Machine Computation of the Complex Fourier Series," *Mathematics of Computation*, 19, 1965.

Doughty, S. "Fourier Series for Unevenly Spaced Data," *Machine Design*, 12 February, 1987, pp. 220–222.

Doughty, S. "Quick Fourier Series," *Access*, May/June, 1989, pp. 22–28.

Franklin, P., *An Introduction to Fourier Methods and the Laplace Transformation*, Dover, 1958.

Kreyszig, E., *Advanced Engineering Mathematics*, 3rd. ed., Wiley, 1972.

Timoshenko, S.P. and Goodier, J.N., *Theory of Elasticity*, 3rd. ed., McGraw-Hill, 1970.

Walker, J.S., *Fast Fourier Transforms*, CRC Press, 1991.

Wylie, C.R., Jr., *Advanced Engineering Mathematics*, 3rd. ed. McGraw-Hill, 1966.

Chapter 12

Rotations

The rotation of physical bodies can be observed every day in the movement of such things as a wheel on an automobile, the blade on a lawn mower, or the impeller of a fan. In the first case of the automobile wheel, translation of the automobile and the wheel is associated with the rotation of the wheel. For the lawn mower, there may or many not be translation associated with the rotation of the blade. In the case of the fan, there is usually no translation at all associated with the rotation of the impeller. Chasle's theorem states that any motion of a rigid body can be considered to consist of a translation and a rotation. The focus of this chapter is on the description of rotation.

The mathematical description of a rotation is somewhat complicated due in part to the fact that there are two, physically distinct, interpretations for the same mathematical description of a rotation. These two interpretations will be referred to here as Point of View 1 and Point of View 2.

Point of View 1

Consider two coordinate systems, X-Y and X'-Y', having a common origin as shown in Figure 1. the coordinate system X'-Y' is rotated an angle θ with respect to X-Y. In that same figure, there is a vector, $\mathbf{r} = x\mathbf{i} + y\mathbf{j} = x'\mathbf{i}' + y'\mathbf{j}'$. If the two coordinate systems are thought of as initially coincident, and the system x-y is then rotated through the angle θ, the question of interest is how the components x', y' are related to the components x, y. Implied in this statement of interest is the fundamental aspect of Point of View 1: *There is only one vector $\mathbf{r}$, and it remains fixed in X-Y while X'-Y' rotates through θ.* This is Point of View 1.

Point of View 2

The second point of view is shown in Figure 2. Point of View 2 says that *there is only one coordinate system, X-Y, and as θ increases from 0, the vector $\mathbf{r}$ moves to become the vector $\mathbf{r}'$.* The prime does not denote a differentiation, but simply a second state. Note that the sense of positive θ is reversed from that in Point of View 1. If the vector $\mathbf{r} = x\mathbf{i} + y\mathbf{j}$ when $\theta = 0$, the usual question of interest is what are the components after the rotation, $\mathbf{r}' = x'\mathbf{i} + y'\mathbf{j}$? Note that this also implies the fundamental aspect of Point of View 2: *There is only one coordinate system and the vector components, and hence the vector, change as θ increases.*

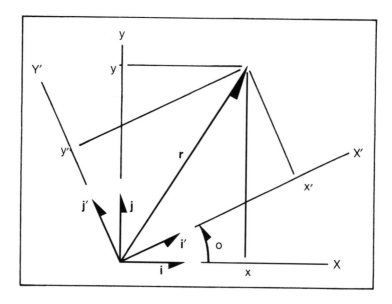

Figure 1: Point of View 1 showing X-Y and X'-Y' coordinate systems.

It should be mentioned again that the mathematical description is the same no matter which of these points of view is adopted. In interpreting the mathematics, it is essential to stay consistently with the same point of view.

Rotations in Two Dimensions

Consider a rotation in the x-y plane, and taking Point of View 1, the appropriate geometry is given by Figure 1. By some elementary right triangle calculations, the primed components can be readily expressed in terms of the unprimed components:

$$
\begin{aligned}
x' &= x\cos\theta + y\sin\theta \\
y' &= -x\sin\theta + y\cos\theta
\end{aligned}
$$

It is often convenient to put this into matrix form as

$$
\left\{ \begin{array}{c} x' \\ y' \end{array} \right\} = \left[\begin{array}{cc} \cos\theta & \sin\theta \\ -\sin\theta & \cos\theta \end{array} \right] \left\{ \begin{array}{c} x \\ y \end{array} \right\}
$$

This may be considered as a strictly two-dimensional result, or it may be expanded to describe a three-dimensional motion involving about the z-axis. For the latter, the z-values are unchanged, so this is expressed as

$$
\left\{ \begin{array}{c} x' \\ y' \\ z' \end{array} \right\} = \left[\begin{array}{ccc} \cos\theta & \sin\theta & 0 \\ -\sin\theta & \cos\theta & 0 \\ 0 & 0 & 1 \end{array} \right] \left\{ \begin{array}{c} x \\ y \\ z \end{array} \right\}
$$

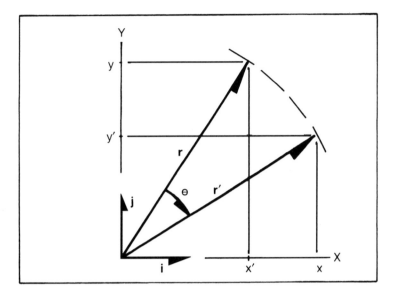

Figure 2: Point of View 2 showing the original vector $\vec{r}$ and the rotated vector $\vec{r'}$.

The third transformation equation is simply $z' = z$, and this is shown by the third row of the transformation matrix that has a 1.0 in the $(3,3)$ position as its only nonzero entry.

If a rotation through angle θ is followed by a rotation through $-\theta$, the net result is that the system is returned to its original position. The rotation through $-\theta$, indicated by the matrix $[T(-\theta)]$, was the inverse of the rotation through θ, indicated by $[T(\theta)]$, and the total rotation is simply the second rotation applied to the result of the first:

$$[T(-\theta)][T(\theta)]\begin{Bmatrix} x \\ y \\ z \end{Bmatrix} = \begin{bmatrix} \cos\theta & -\sin\theta & 0 \\ \sin\theta & \cos\theta & 0 \\ 0 & 0 & 1 \end{bmatrix}\begin{bmatrix} \cos\theta & \sin\theta & 0 \\ -\sin\theta & \cos\theta & 0 \\ 0 & 0 & 1 \end{bmatrix}\begin{Bmatrix} x \\ y \\ z \end{Bmatrix}$$

$$= \begin{bmatrix} \cos^2\theta + \sin^2\theta & \cos\theta\sin\theta - \sin\theta\cos\theta & 0 \\ \sin\theta\cos\theta - \cos\theta\sin\theta & \sin^2\theta + \cos^2\theta & 0 \\ 0 & 0 & 1 \end{bmatrix}\begin{Bmatrix} x \\ y \\ z \end{Bmatrix}$$

$$= \begin{bmatrix} 1 & 0 & 0 \\ 0 & 1 & 0 \\ 0 & 0 & 1 \end{bmatrix}\begin{Bmatrix} x \\ y \\ z \end{Bmatrix}$$

Note that the second rotation matrix, $[T(-\theta)]$ is simply the transpose of the first rotation, $[T(\theta)]$. This demonstrates an important property of a rotational transformation matrix: *The inverse of a rotation matrix must be the transpose of the rotation matrix.* Although this has been shown only for a rotation about the z-axis, it is true in general as well.

Proper and Improper Rotations

The statement that $[T(\theta)]^{-1} = [T(\theta)]^t$ must be true for any rotation, but it is not sufficient to assure that a particular matrix represents a rotation. From above, for the two successive rotations

$$[T(-\theta)][T(\theta)] = [T(\theta)]^{-1}[T(\theta)] = [T(\theta)]^t[T(\theta)] = [I] = \text{ identity matrix}$$

The determinant of a transpose is the same as the determinant of the original matrix, so taking the determinant of both sides of this relation

$$\left| [T(\theta)]^t[T(\theta)] \right| = \|[T(\theta)]\|^2 = \|[I]\| = +1$$

Thus, from the requirement that the inverse of the transformation must be the transpose of the transpose, there follows that the determinant of the transformation must be ± 1. If the determinant is $+1$, then the matrix represents a *proper* rotation. An inspection of the transformation matrix presented above for rotation about the z-axis shows that its determinant is in fact $+1$. What is the significance of a -1 determinant value? If the determinant is -1, the transformation is said to be an *improper* rotation, but that does not shed very much light on what happens. An improper rotation turns space "inside out," that is it transforms a right-handed coordinate system into a left-handed system and vice versa. Such a displacement cannot happen in terms of physical solid bodies, therefore, in the physical world, improper rotations do not exist. Thus, for $[T(\theta)]$ to be a proper rotation, there are two requirements:

- $[T(\theta)]^{-1}$ must be equal to $[T(\theta)]^t$ and

- $\|[T(\theta)]\|$ must be equal to $+1$.

Although these properties have been demonstrated only for rotations about the z-axis, they are true in general for any rotation matrix.

Rotations in Three Dimensions

To begin the discussion of rotations in three dimensions, again let us adopt Point of View 1 and consider Figure 3. In this discussion, there will be need for many cosine expressions, and the notation that will be used is

$$\begin{aligned} c_{p,q} &= \cos(\text{angle between lines p and q}) \\ &= c_{q,p} \end{aligned}$$

142

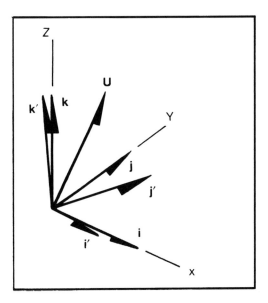

Figure 3: Primed unit vectors resulting from a rotation.

simply to shorten the notation. Note that it is permissible to interchange the order of the indices because the cosine is an even function and is therefore unaffected by a change in the sign of the argument.

Direction Cosine Matrix

Before transforming vector components, the transformation of the unit vectors themselves must be examined. Consider the process of projecting each of the primed unit vectors onto the unprimed axes and writing them in terms of the unprimed unit vectors:

$$
\begin{aligned}
\mathbf{i}' &= c_{x',x}\mathbf{i} + c_{x',y}\mathbf{j} + c_{x',z}\mathbf{k} \\
\mathbf{j}' &= c_{y',x}\mathbf{i} + c_{y',y}\mathbf{j} + c_{y',z}\mathbf{k} \\
\mathbf{k}' &= c_{z',x}\mathbf{i} + c_{z',y}\mathbf{j} + c_{z',z}\mathbf{k}
\end{aligned}
$$

This may be cast in matrix form as

$$
\left\{ \begin{array}{c} \mathbf{i}' \\ \mathbf{j}' \\ \mathbf{k}' \end{array} \right\} = \left[\begin{array}{ccc} c_{x',x} & c_{x',y} & c_{x',z} \\ c_{y',x} & c_{y',y} & c_{y',z} \\ c_{z',x} & c_{z',y} & c_{z',z} \end{array} \right] \left\{ \begin{array}{c} \mathbf{i} \\ \mathbf{j} \\ \mathbf{k} \end{array} \right\} = [C] \left\{ \begin{array}{c} \mathbf{i} \\ \mathbf{j} \\ \mathbf{k} \end{array} \right\}
$$

The coefficient matrix is composed entirely of the nine direction cosines relating the primed and unprimed axes. Note that $[C]$ is exactly equivalent to the matrix previously denoted as $[T]$, although $[T]$ was not stated entirely in terms of cosines. If we consider going the other way, projecting each of the unprimed unit vectors onto the primed axes, the relations are

$$
\begin{aligned}
\mathbf{i} &= c_{x,x'}\mathbf{i}' + c_{x,y'}\mathbf{j}' + c_{x,z'}\mathbf{k}' \\
\mathbf{j} &= c_{y,x'}\mathbf{i}' + c_{y,y'}\mathbf{j}' + c_{y,z'}\mathbf{k}' \\
\mathbf{k} &= c_{z,x'}\mathbf{i}' + c_{z,y'}\mathbf{j}' + c_{z,z'}\mathbf{k}'
\end{aligned}
$$

or

$$
\left\{ \begin{array}{c} \mathbf{i} \\ \mathbf{j} \\ \mathbf{k} \end{array} \right\}
=
\left[\begin{array}{ccc}
c_{x',x} & c_{x,y'} & c_{x,z'} \\
c_{y,x'} & c_{y,y'} & c_{y,z'} \\
c_{z,x'} & c_{z,y'} & c_{z,z'}
\end{array} \right]
\left\{ \begin{array}{c} \mathbf{i}' \\ \mathbf{j}' \\ \mathbf{k}' \end{array} \right\}
= [C]^t \left\{ \begin{array}{c} \mathbf{i}' \\ \mathbf{j}' \\ \mathbf{k}' \end{array} \right\}
$$

Note that the transformation in this direction is seen immediately as the transpose of the matrix representing the previous direction. Now that we know how the unit vectors are transformed, how are the vector components themselves transformed?

Consider a vector $\mathbf{r}$,

$$
\mathbf{r} = (x, y, z) \left\{ \begin{array}{c} \mathbf{i} \\ \mathbf{j} \\ \mathbf{k} \end{array} \right\}
= (x', y', z') \left\{ \begin{array}{c} \mathbf{i}' \\ \mathbf{j}' \\ \mathbf{k}' \end{array} \right\}
= (x', y', z') [C] \left\{ \begin{array}{c} \mathbf{i} \\ \mathbf{j} \\ \mathbf{k} \end{array} \right\}
$$

Since there is no division of vectors, this cannot be solved simply by dividing out the vectors. It can, however, be solved as follows:

$$
\left\langle (x, y, z) - (x', y', z') [C] \right\rangle \left\{ \begin{array}{c} \mathbf{i} \\ \mathbf{j} \\ \mathbf{k} \end{array} \right\}
= \left\{ \begin{array}{c} \mathbf{0} \\ \mathbf{0} \\ \mathbf{0} \end{array} \right\}
$$

from which we may conclude that the quantity in the $\langle\,\rangle$ must be zero,

$$
(x, y, z) = (x', y', z') [C]
$$

or, taking the transpose to put the vectors in column form as usual and then solving for the primed components,

$$
\left\{ \begin{array}{c} x' \\ y' \\ z' \end{array} \right\}
= [C] \left\{ \begin{array}{c} x \\ y \\ z \end{array} \right\}
$$

As expected, this is the same transformation law as was obtained for the unit vectors themselves. Thus the matrix of direction cosines is the transformation matrix from the unprimed to primed coordinates.

Transformation in Terms of Axis and Angle

It is one thing to be able to say that a particular matrix represents a rotation, but it is quite a different matter to say how much rotation and about what axis the rotation is made. This topic is addressed next, and in that discussion, it is

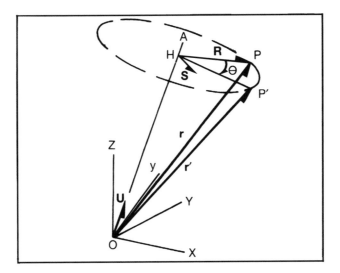

Figure 4: Vector r rotated to r' by rotation matrix $[C]$.

convenient to take Point of View 2, that there is only one coordinate system and that the vector itself moves. Such a situation is shown in Figure 4.

The line from the origin through point A is the axis of rotation, and the rotation θ will move the vector $\mathbf{r}$ toward the user to the position $\mathbf{r'}$. The axis is initially specified by giving the coordinates of a point on the axis, and those coordinates serve as direction numbers for a line through that point and the origin. From the direction numbers, the direction cosines are determined in the usual fashion, and it is assumed that they are known, that is,

$$\begin{aligned} \cos\left(\overline{OA}, X\right) &= L \\ \cos\left(\overline{OA}, Y\right) &= M \\ \cos\left(\overline{OA}, Z\right) &= N \end{aligned}$$

are all known. If $\mathbf{U}$ denotes a unit vector along the axis, it is

$$\mathbf{U} = L\mathbf{i} + M\mathbf{j} + N\mathbf{k}$$

The length $\overline{OH}$ is the projection of the vector $\mathbf{r} = col\,(x, y, z)$ with tip at point P onto the axis of rotation. The vector $\mathbf{R}$ extends from point H to point P:

$$\mathbf{R} = \mathbf{r} - \overline{OH}\,\mathbf{U} = \mathbf{r} - (\mathbf{r} \cdot \mathbf{U})\,\mathbf{U}$$

The vector $\mathbf{R}$ is not a unit vector. Let R denote the magnitude of $\mathbf{R}$.

The vector $\mathbf{r}$ and the unit vector $\mathbf{U}$ define a plane. If $\mathbf{S}$ is a unit vector normal to that plane, it may be determined by

$$\mathbf{S} = \frac{-\mathbf{U} \times \mathbf{R}}{|\mathbf{U} \times \mathbf{R}|} = \frac{-\mathbf{U} \times \left(\mathbf{r} - \overline{OH}\,\mathbf{U}\right)}{R} = \frac{-\mathbf{U} \times \mathbf{r}}{R}$$

From Figure 4, it is evident that the new vector $\mathbf{r}'$ is expressible as

$$\mathbf{r}' = \overline{OH}\,\mathbf{U} + \mathbf{R}\cos\theta + \mathbf{S}\,R\sin\theta$$

where the first term gives the component along the axis of rotation, the second terms gives the component along $\mathbf{R}$, and the third term gives the component normal to both $\mathbf{U}$ and $\mathbf{R}$. Now consider the further development of each part of this expression:

$$
\begin{aligned}
\mathbf{r}' &= \overline{OH}\,\mathbf{U} + R\cos\theta + \mathbf{S}\,R\cos\theta \\
&= \overline{OH}\,\mathbf{U} + \left(\mathbf{r} - \overline{OH}\,\mathbf{U}\right)\cos\theta + \frac{-\mathbf{U}\times\mathbf{r}}{R}R\sin\theta \\
&= \mathbf{r}\cos\theta + \overline{OH}\,\mathbf{U}\,(1-\cos\theta) + \mathbf{r}\times\mathbf{U}\sin\theta
\end{aligned}
$$

Consider the first term

$$
\begin{aligned}
\overline{OH}\mathbf{U} &= (\mathbf{r}\cdot\mathbf{U})\,\mathbf{U} \\
&= \mathbf{U}\mathbf{U}\cdot\mathbf{r} = \{U\}\,(U)\,\{r\} \\
&= \left\{\begin{array}{c} L \\ M \\ N \end{array}\right\} (L, M, N) \left\{\begin{array}{c} x \\ y \\ z \end{array}\right\} \\
&= \begin{bmatrix} L^2 & LM & LN \\ LM & M^2 & MN \\ LN & MN & N^2 \end{bmatrix} \left\{\begin{array}{c} x \\ y \\ z \end{array}\right\}
\end{aligned}
$$

and a part of the last term

$$
\begin{aligned}
\mathbf{r}\times\mathbf{U} &= -\mathbf{U}\times\mathbf{r} \\
&= -[Ux]\left\{\begin{array}{c} x \\ y \\ z \end{array}\right\} \\
&= \begin{bmatrix} 0 & N & -M \\ -N & 0 & L \\ M & -L & 0 \end{bmatrix} \left\{\begin{array}{c} x \\ y \\ z \end{array}\right\}
\end{aligned}
$$

With all of the pieces reassembled, the complete result is

$$
\begin{aligned}
\left\{\begin{array}{c} x' \\ y' \\ z' \end{array}\right\} = {}& \cos\theta \begin{bmatrix} 1 & 0 & 0 \\ 0 & 1 & 0 \\ 0 & 0 & 1 \end{bmatrix} \left\{\begin{array}{c} x \\ y \\ z \end{array}\right\} \\
&+ (1-\cos\theta) \begin{bmatrix} L^2 & LM & LN \\ LM & M^2 & MN \\ LN & MN & N^2 \end{bmatrix} \left\{\begin{array}{c} x \\ y \\ z \end{array}\right\} \\
&+ \sin\theta \begin{bmatrix} 0 & N & -M \\ -N & 0 & L \\ M & -L & 0 \end{bmatrix} \left\{\begin{array}{c} x \\ y \\ z \end{array}\right\}
\end{aligned}
$$

or

$$[C] = \cos\theta \begin{bmatrix} 1 & 0 & 0 \\ 0 & 1 & 0 \\ 0 & 0 & 1 \end{bmatrix} + (1 - \cos\theta) \begin{bmatrix} L^2 & LM & LN \\ LM & M^2 & MN \\ LN & MN & N^2 \end{bmatrix} + \sin\theta \begin{bmatrix} 0 & N & -M \\ -N & 0 & L \\ M & -L & 0 \end{bmatrix}$$

$$\begin{Bmatrix} x' \\ y' \\ z' \end{Bmatrix} = [C] \begin{Bmatrix} x \\ y \\ z \end{Bmatrix}$$

If the axis and the angle are specified, this development shows how the rotation matrix is expressed in terms of those parameters.

Axis and Angle From Direction Cosine Matrix

In the previous section, the problem of determining a direction cosine matrix for a specified rotation axis and angle was addressed. Here the problem is reversed: Given a direction cosine matrix, what are the axis of rotation and the angle of rotation?

Beginning with the given direction cosine matrix, $[C]$,

$$[C] = \begin{bmatrix} c_{x',x} & c_{x',y} & c_{x',z} \\ c_{y',x} & c_{y',y} & c_{y',z} \\ c_{z',x} & c_{z',y} & c_{z',z} \end{bmatrix}$$

compute the difference $[C] - [C]^t$

$$[C] - [C]^t = \begin{bmatrix} 0 & c_{x',y} - c_{y',x} & c_{x',z} - c_{z',x} \\ c_{y',x} - c_{x',y} & 0 & c_{y',z} - c_{z',y} \\ c_{z',x} - c_{x',z} & c_{z',y} - c_{y',z} & 0 \end{bmatrix} = \begin{bmatrix} 0 & \tilde{N} & -\tilde{M} \\ -\tilde{N} & 0 & \tilde{L} \\ \tilde{M} & -\tilde{L} & 0 \end{bmatrix}$$

Comparing this with the same quantity expressed in terms of the results of the previous section

$$[C] - [C]^t = 2\sin\theta \begin{bmatrix} 0 & N & -M \\ -N & 0 & L \\ M & -L & 0 \end{bmatrix}$$

Equating these two expressions and then extracting the three equations represented there, the results are

$$\begin{aligned} 2\sin\theta\, L &= \tilde{L} \\ 2\sin\theta\, M &= \tilde{M} \\ 2\sin\theta\, N &= \tilde{N} \end{aligned}$$

If each of these three equations is squared and the results added, the sum is

$$4\sin^2\theta \left(L^2 + M^2 + N^2\right) = 4\sin^2\theta = \tilde{L}^2 + \tilde{M}^2 + \tilde{N}^2 = \tilde{R}^2$$

147

Taking the square root, an expression relating the angle θ and the elements of $[C] - [C]^t$ results:

$$\sin\theta = \pm\frac{1}{2}\sqrt{\tilde{L}^2 + \tilde{M}^2 + \tilde{N}^2}$$

and from this result, the direction cosines for the axis are determined,

$$L = \frac{\pm\tilde{L}}{\sqrt{\tilde{L}^2 + \tilde{M}^2 + \tilde{N}^2}}$$

$$M = \frac{\pm\tilde{M}}{\sqrt{\tilde{L}^2 + \tilde{M}^2 + \tilde{N}^2}}$$

$$N = \frac{\pm\tilde{N}}{\sqrt{\tilde{L}^2 + \tilde{M}^2 + \tilde{N}^2}}$$

One other piece of information is available from the main diagonal of the direction cosine matrix and the expression for the direction cosine matrix in terms of the axis of rotation and the angle of rotation:

$$c_{xx'} = \cos\theta + (1 - \cos\theta)\,L^2$$

$$c_{yy'} = \cos\theta + (1 - \cos\theta)\,M^2$$

$$c_{zz'} = \cos\theta + (1 - \cos\theta)\,N^2$$

If these are added, the squares of the direction cosines sum to 1.0, and the result is solvable for the cosine of the rotation angle:

$$\cos\theta = \frac{1}{2}\left(c_{xx'} + c_{yy'} + c_{zz'} - 1\right)$$

The expression for the cosine of the rotation angle is unambiguous with respect to sign, but the expressions for the sine of the rotation angle and for the direction cosines of the axis of rotation all involve a double sign. Consider first the case where the positive sign is taken at all locations involving the double sign, remembering that $\tilde{L}, \tilde{M}$, and $\tilde{N}$ each still carry a sign. This case defines a particular axis and angle of rotation. If now the negative sign is taken at all double sign locations, the direction cosines all change sign as compared to the previous case, indicating the same axis of rotation except that the unit vector points in the opposite direction along that axis of rotation. The expression for the sine of the rotation angle also changes signs, indicating that the sense of rotation is reversed. The result is that the rotation is the same as that previously described. A rotation of amount $+\theta$ described by the unit vector $\mathbf{U}$ is exactly equivalent to a rotation of amount $-\theta$ about an axis described by $-\mathbf{U}$. Therefore all of the double signs can be dropped, and the results are as summarized here:

$$L = \frac{\tilde{L}}{\sqrt{\tilde{L}^2 + \tilde{M}^2 + \tilde{N}^2}}$$

$$M = \frac{\tilde{M}}{\sqrt{\tilde{L}^2 + \tilde{M}^2 + \tilde{N}^2}}$$

$$N = \frac{\tilde{N}}{\sqrt{\tilde{L}^2 + \tilde{M}^2 + \tilde{N}^2}}$$

$$\sin\theta = \frac{1}{2}\sqrt{\tilde{L}^2 + \tilde{M}^2 + \tilde{N}^2}$$

$$\cos\theta = \frac{1}{2}\left(c_{xx'} + c_{yy'} + c_{zz'} - 1\right)$$

The two argument arctangent function (Appendix 1) is useful at this point to extract the actual rotation angle, θ.

References

Crandall, S.H., Karnopp, D.C., Kurtz, E.F., Jr., and Pridmore-Brown, D.C., *Dynamics of Mechanical and Electromechanical Systems*, McGraw-Hill, 1968.

Goldstein, H., *Classical Mechanics*, Addison-Wesley, 1959.

Synge, J.L., "Classical Dynamics," *Encyclopedia of Physics*, S. Flugge, ed., vol. III/1, Springer Verlag, 1960.

Appendix 1

Two Argument Arctangent Function

The arctangent operation is required for the conversion from rectangular to polar coordinates. Because of the multivalued nature of the arctangent, it is not strictly a function. Most programming languages use the principal value of the arctangent which has a unique value, lying in either the first or fourth quadrants. This is, however, not sufficient for the polar to rectangular conversion, because for points in the left half plane, the argument angle will lie in either the second or third quadrants. Fortran provides a two argument arctangent operation which resolves this difficulty, but most other languages do not offer this option. The purpose for the subroutine presented below is to answer this need. It is thought that this code is largely self-explanatory, but a few comments are offered after the listing.

Two Argument Arctangent

```
sub at2(x,y,theta)
    ! A two argument arctangent subroutine
    ! Inputs are x and y; output is angle theta
    if x>0 then
        theta=atn(y/x)
    else if x=0 then
        if y>0 then theta=pi/2
        if y=0 then theta=0
        if y<0 then theta=-pi/2
    else if x<0 then
        s=+1
        if y<0 then s=-1
        theta=atn(y/x)+s*pi
    end if
end sub
```

The value returned for the angle is in radian measure, as is the usual form for angular measure in computer programs. This angle will be in the interval $-\pi < \theta \le \pi$. For the case where both $x = 0$ and $y = 0$, zero is arbitrarily assigned to the angle; in fact, the angle is not defined in this case. Note also that this routine assumes an internally defined value for π. If such a value is not available, then an additional line defining π is required.

150

Appendix 2

Two Dimensional Representation of Space Curves

Many physical problems require consideration of three dimensional geometries, but we are generally limited to two dimensions anytime we want to draw pictures. This leads to efforts to provide the third dimension by use of some sort of perspective. In terms of engineering graphics, there are several standard methods, including isometric and oblique views. For present purposes, it is assumed that the data consists of three lists, $x(s)$, $y(s)$, $z(s)$, where each list contains N values. The approach is to determine a method for correctly locating a typical point, say $x(s_i)$, $y(s_i)$, $z(s_i)$, and then to apply that method to all points. For convenience, the subscript i is dropped at this point.

The drawing space will be described by two coordinates, U and V, and for convenience they are located at the far left lower corner of the space, so that all values for U and V are positive. The origin of coordinates for the three-dimensional system will be located at the point $U = U_o$, $V = V_o$ in the drawing space. The selection of U_o and V_o is entirely the user's choice, and is often something to be adjusted until the final result is suitable. The location of the origin is shown in Figure 1.

In order to achieve the necessary perspective, the X and Y axes are located at angles A and B, with respect to the horizontal, as shown in Figure 1. As will the location of the origin, the choice of values for A and B is entirely a matter of the user's choice. With A and B defined as shown in Figure 1, the perspective can be adjusted to include the two standard choices,

A	B	Drawing Type
$\frac{\pi}{3} = 60^o$	$\frac{\pi}{3} = 60^o$	Isometric
0	$\frac{\pi}{2} = 45^o$	Oblique

While these standard types are available, there is no need to be confined to them. It is generally desirable to adjust A and B iteratively until the picture shows the features important to the particular plot, rather than being restricted to the standard choices.

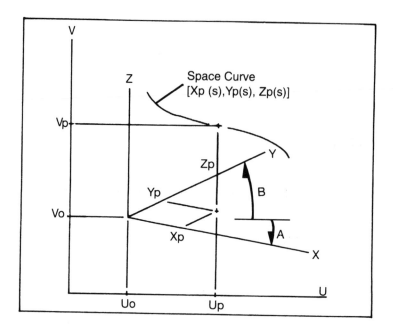

Figure 1: Two dimensional representation of a space curve.

For a typical point on the space curve, the plotting location is given by the following equations:

$$U_p = U_o + x_p(s)\cos A + y_p(s)\cos B$$
$$V_p = V_o - x_p(s)\sin A + y_p(s)\sin B + z_p(s)$$

It is often useful to make a short subroutine including these equations, and then for each point to be plotted, the subroutine is called once to determine the values of U_p, V_p.

Appendix 3

Programs on Diskette

The disk that accompanies this book includes all of the programs included in the book as well as several others. The purpose of this appendix is to provide a brief description of each of these programs and to connect it with the appropriate section of the text. The programs are considered here in groups corresponding to the chapters from which they are drawn.

The extensions used with the file names indicate the type of code in the file. Four extensions are used:

.Tru — Indicates a complete program in True BASIC source code

.Sub — Indicates a complete subroutine in True BASIC source code

.Frg — Indicates a program fragment in True BASIC source code

.Exe — Indicates a compiled MS-DOS program based on the file with the extension .Tru

The program fragments cannot be used without further work by the user. They are intended to show the structure of various algorithms.

All of the subroutines that are listed here are of the type called *external subroutines* in True BASIC. These subroutines belong outside the main program, after the end statement, in contrast to the internal subroutines used in older versions of BASIC. External subroutines are called by name with an argument list. This means that the only communication between the main routine and the subroutine is through the argument list; names are not shared between the two routines. Some of the complete program examples also make use of internal subroutines as well.

In a few of the listings the user will find lines that end with &, followed by a line that begins with &. This is a line continuation and is to be interpreted as if the second line were appended directly to the end of the first line. This is done for readability only. It has no impact on the calculation at all.

Chapter 2 — Matrices

MatProd.Frg

This short program fragment shows the necessary loop structure to carry out a matrix product if the language used does not support single line matrix commands. It is complete in itself, but before it is used the array names must be adjusted to fit the application.

Matin.Sub

This subroutine has three uses:

- Determining the solution for a system of linear equations

- Determining the inverse of a nonsingular square matrix

- Determining the determinant of a square matrix

This subroutine is useful principally in languages that do not support single-line matrix operations.

MatinDm1.Tru

This program is an example of the **Matin** subroutine used to solve a system of three linear, simultaneous equations. For this particular example the system shown is

$$
\begin{aligned}
2x + 3y + 7z &= 15 \\
x - 4y - 8z &= 12 \\
-3x + y + 2z &= 0
\end{aligned}
$$

The program initially fills the arrays representing the coefficient matrix and the right-side vector, and then calls the subroutine for the solution. Note that the coefficient matrix is dimensioned 4×3 as will be needed in the augmented form. The right-side vector must be defined as a two-dimensional array since the subroutine allows for the possibility of multiple-right side vectors. The solution found for this system is

$$
\begin{aligned}
x &= -1.0909091 \\
y &= -57.272727 \\
z &= 27.0
\end{aligned}
$$

$$\text{Determinant} \;=\; -11$$

The solution is checked by the program by direct numerical substitution to calculate the residual for each equation.

MatinDm2.Tru

This program demonstrates the use of the `Matin` subroutine for matrix inversion. The matrix to be inverted is called $[B]$, a 3×3 matrix. It is the matrix of coefficients from the program `MatinDm1.Tru`. The original dimension for $[B]$ is established as 3×6 to allow room for the augmented form that will be required. Note also that the number of right side vectors, `naug`, is set to three as is consistent with the augmented matrix. The computed inverse is

$$[B]^{-1} = \begin{bmatrix} 0 & -9.0909091\,(10^{-2}) & -0.36363636 \\ -2 & -2.2727273 & -2.0909091 \\ 1 & 1 & 1 \end{bmatrix}$$

The final results listed on the screen are the original matrix $[B]$, the matrix $[B]^{-1}$, and the computed product $[B][B]^{-1}$. This last should be an identity matrix, and it almost is.

Chapter 3 — Vectors

ACross.Sub

This small subroutine transforms a vector into a square matrix representing the first factor of a vector cross product.

DotCross.Tru

This subroutine demonstrates vector dot and cross products performed in the computer. The subroutine `ACross` is used in the cross product calculation. This program also includes a subroutine `DotProd` for the calculation of a dot product.

Chapter 5 — Smoothing

Smooth35.Sub

This subroutine provides smoothing based on a cubic curve that is a least squares fit to five data points. It is taken from the Examples section of Chapter 5.

Chapter 6 — Interpolation and Differentiation

LagrPoly.Sub

This subroutine provides for interpolation and differentiation of a tabulated function based on the Lagrange interpolating polynomial. It assumes that the polynomial is to be fitted to data with indices p to q in the tabulation, and the evaluation point is denoted as xx.

LagrPoly.Tru

This program is an application of the `LagrPoly` subroutine, specifically designed to demonstrate the capability of the subroutine to handle both evenly and unevenly spaced data. Both interpolation and differentiation are demonstrated. The necessary routines to search for the required data subset (index values p to q) are also provided.

NewtIntp.Tru

This program demonstrates the use of the Newton Divided Difference technique for interpolation of evenly or unevenly spaced data. The necessary interval locating steps are included.

CentDiff.Tru

This program determines the coefficients for a Central Difference Approximation with evenly spaced data. The user enters the number of data points to be used in each central difference calculation, and the program produces the coefficients for all of the derivatives that can be approximated.

Chapter 7 — Applications of Green's Theorem

PlnArea.Tru

This program computes the area, centroidal location, and area moments of inertia with respect to the coordinate axes for any planar area. The boundary curve defining the area is described by a sequence of nodes that encompass the area in a counter-clockwise direction. Curved sections of the boundary are approximated by a sequence of straight line segments, and the validity of the results depends on how well this approximates the true boundary.

As a test problem for this program, consider a plane figure in the form of a square but with a triangle cut away, as described by the following sequence of points: $(-2, -2)$, $(3, -2)$, $(0, 1)$, $(3, 0)$, $(3, 3)$, $(-2, 3)$. The computed results for this figure are

$$
\begin{aligned}
A &= 22 \\
X_c &= 0.2954545 \\
Y_c &= 0.6136364 \\
I_{xx} &= 56.833333 \\
I_{yy} &= 44.833333
\end{aligned}
$$

These values may be verified by a hand calculation that will also serve to show the usefulness of the program.

SolRev.Tru

This program computes surface and volume properties for a body of revolution. The curve defining the body of revolution is approximated by a series of straight line segments, using nodes that traverse the boundary in a counterclockwise direction.

As a test problem for this program, consider the right circular cone generated by revolving the $r-z$ plane triangle $(0,0)$, $(2,3)$, $(0,3)$ about the z-axis. Some of the results for this case are given below and can of course be check by hand calculation:

$$
\begin{aligned}
\text{Surface Area} &= 35.2207 \\
\text{Volume} &= 12.5664 \\
\bar{z} &= 2.25 \\
I_{rr} &= 75.39822 \\
I_{zz} &= 15.07964
\end{aligned}
$$

Chapter 8 — Roots of a Single Equation

NewtonEx.Tru

This program is an example of the application of Newton's Root Finding Method to the problem of determining the natural frequencies of flexible, massive beam supported on two springs. The solution details of the problem have been developed in the text as an example there. The function for which roots were required in that example was

$$
\begin{aligned}
f(\lambda) = {} & -2\cos\lambda L \cosh\lambda L + \cos^2\lambda L + 2\left(\cos\lambda L\right)k_2\sinh\lambda L \\
& + \cosh^2\lambda L - 2\left(\cosh\lambda L\right)k_2\sin\lambda L + 2k_1\sinh\lambda L\cos\lambda L \\
& + 4k_1\left(\sinh\lambda L\right)k_2\sin\lambda L + \sin^2\lambda L - 2\left(\sin\lambda L\right)k_1\cosh\lambda L \\
& - \sinh^2\lambda L
\end{aligned}
$$

NewtnEx2.Tru

This is an example of Newton's method applied to a much simpler equation than that considered previously. The problem here is to located a root in the interval $1.25 \leq x \leq 2.25$ for the equation

$$\sin x + \cos x = 0.5$$

With a starting estimate of 1.75, the solution converges to 1.9948274 with a residual of $-9.12\,(10^{-10})$. Other starting values in the interval give comparable results.

Bairstow.Tru

This program demonstrates Bairstow's method for finding the real and complex roots of a polynomial with real coefficients. The example problem developed in the text is the determination of the roots of the equation

$$x^4 - 16 = 0$$

for which the roots are readily observed to be $+2 + j0$, $-2 + j0$, $0 + j2$, and $0 - j2$.

Bairsto2.Tru

This is a second example of Bairstow's method for the roots of a polynomial. The polynomial considered here is

$$x^3 - 5x^2 + 17x - 13 = 0$$

for which the roots are known to be $x = 2 + j3$, $x = 2 - j3$, $x = 1 + j0$. The only changes required are in the value for **n** and the list of polynomial coefficients.

Chapter 9 — Systems of Nonlinear Equations

NewtRaph.Tru

This program demonstrates the application of the Newton-Raphson Method for the solution of systems of nonlinear equations. The particular problem solved in this example is

$$
\begin{aligned}
x^2 + 3xy - y &= 0 \\
\sin x + y &= 4
\end{aligned}
$$

solved for a root in the domain $0 \leq x \leq 1$, $0 \leq y \leq 5$.

NewtRap2.Tru

This is a second example of the Newton-Raphson method, this time applied to a system of three equations. There are modifications required throughout the program, including changes to (1) the function evaluation subroutine, (2) the Jacobian evaluation, (3) the calculation of the squared sum of residuals (`fsq`), (4) the calculation of the squared sum of the adjustments (`dsq`), and, of course, the dimension statements. The problem is to find a root of the system of equations

$$\begin{aligned} x + y + z &= 1 \\ xy + z &= -11 \\ xyz &= -80 \end{aligned}$$

near the point $(4, -4, 4)$. After six iterations the solution converges to

$$\begin{aligned} x &= 2.472136 \\ y &= -6.472136 \\ z &= 5.0 \end{aligned}$$

Chapter 10 — Ordinary Differential Equations

RKF5.Frg

This fragment gives the structure of the Runge-Kutta-Feldberg 5th order algorithm as applied to a single, first-order ordinary differential equation. The details of a particular problem are missing and marked by the ellipsis (...). The details for any particular problem may be substituted with very little effort, so that this fragment represents a large part of the effort required to solve any problem.

RKF5.Tru

This is an application of the RKF5 fragment to solve the first-order initial condition problem

$$\dot{y} = -yt \qquad\qquad y(0) = 2$$

The exact solution for this problem is readily found by separating the variables and integrating, and it is

$$y(t) = y(0)\, e^{-t^2/2}$$

This solution has been incorporated into the code for comparison. The tabulated results are listed from left to right as time, computed solution, exact solution, computed derivative, and step size. The last has been included simply to show the variable step size at work. With the allowable error at each step set at 10^{-5}, the program requires 537 steps to reach $t = 2$. At the last point, the error between the exact and computed solutions is 0.057.

RK4Sclr.Frg

This fragment gives the application of the fourth-order Runge-Kutta algorithm in the form applicable to a single second order ordinary differential equation. The details of a particular problem are missing and marked by the ellipsis (...). Completion of the program for any particular problem is a simple matter of replacing the ellipsis items with the expressions appropriate to the particular problem.

RK4Sclr.Tru

This program is an application of the RK4Sclr fragment used to solve the differential equation

$$\ddot{y} + 2\zeta\omega_n\dot{y} + \omega_n^2 y = 0 \qquad y(0) = 1, \quad \dot{y}(0) = 1$$

The values $\zeta = 0.2$ and $\omega_n = 2.0$ have been included directly in the second derivative subroutine. A plotting subroutine is included also that will plot the computed solutions and label them.

RK4Vect.Frg

This fragment gives the application of the fourth-order Runge-Kutta algorithm in the form suited to a system of second-order differential equations. As before, the particular problem details are missing and marked with the ellipsis (...).

Chapter 11 — Fourier Series

FSCoef.Sub

This subroutine employs the sequence calculation in the determination of the Fourier series coefficients for an evenly spaced, tabulated function f(). The argument list includes the array capF(), which is a scratch vector used in the calculation. The Fourier series coefficients are retuned in the argument arrays a() and b().

SerEval.Sub

This subroutine employs the sequence calculation for summing the series, the process of evaluating a Fourier series at a particular argument. It makes no difference at all how the series coefficients were first obtained; the series can be summed using this subroutine. The evaluation point is denoted as x, and the period of the function is denoted as L in the argument list. The Fourier series coefficients are passed in the arrays a() and b(), and the additional arrays capA() and capB() are used as scratch vectors in the calculation.

UFSCoef.Sub

This subroutine employs the polynomial interpolation and Simpson's rule for the evaluation of the integrals appearing in the Fourier series coefficient definitions.

subroutine `FSCoef` is recommended).

FourDem1.Tru

This is a demonstration of `FSCoef` and `SerEval` applied to a wave form for which the Fourier coefficients are known exactly. The values of the Fourier coefficients are specified and used in the creation of the tabulated function that is then sent to `FSCoef`. The returned coefficients compare very favorably with those used to create the function. Then the computed Fourier series coefficients are used to reconstruct the function using the subroutine `SerEval`. The user is asked to specify N_2, the number of orders to be used in the reconstruction. It is suggested that N_2 be tried as 1, 2, 3, 4, ... in successive runs to see how this changes the reconstruction.

FourDem2.Tru

In this application of the subroutines `FSCoef` and `SerEval`, the original wave form is one cycle of a saw tooth, $y = 2x/L$. This function is an odd function, so all of the cosine coefficients of order higher than zero are zero. The first cosine coefficient, a_0, is the average value and it is calculated as 1.0. The sine coefficients for all orders greater than zero are nonzero. This function is discontinuous and this leads to (1) a truly infinite series, and (2) an oscillation in the reconstructed wave form near the points of discontinuity. This oscillation is called Gibb's phenomenon and is clearly evident in the plot of the reconstructed wave form. The user is asked to specify N_2, the number of orders to be used in the reconstruction. A little experimentation will show that including more orders improves the representation. It should be noted, however, that the Gibb's phenomenon will not be eliminated for any finite number of orders.

Appendix 1 — Two Argument Arctangent

At2.Sub

This little subroutine provides a two argument arctangent function. The coordinates of a point for which an arctangent is to be evaluated are passed as x and y. The arctangent value in the proper quadrant is returned as theta.

161

Index